Élisée Reclus

The Ocean, Atmosphere and Life

Being the second series of a descriptive history of the phenomena of the life of the

globe

Élisée Reclus

The Ocean, Atmosphere and Life
Being the second series of a descriptive history of the phenomena of the life of the globe

ISBN/EAN: 9783337036553

Printed in Europe, USA, Canada, Australia, Japan

Cover: Foto ©ninafisch / pixelio.de

More available books at **www.hansebooks.com**

THE OCEAN, ATMOSPHERE AND LIFE

Being the Second Series of

A DESCRIPTIVE HISTORY OF THE

PHENOMENA OF THE LIFE OF THE GLOBE

BY ÉLISÉE RECLUS

Translated by the late B. B. WOODWARD, M.A.

AND

Edited by HENRY WOODWARD, F.R.S., OF THE BRITISH MUSEUM

Illustrated by

TWO HUNDRED MAPS OR FIGURES INTERCALATED IN THE TEXT
AND TWENTY-SEVEN MAPS PRINTED IN COLOURS

SECTION I.

LONDON
CHAPMAN AND HALL, 193, PICCADILLY
1873

JOHN CHILDS AND SON, PRINTERS.

CONTENTS.

PART I.—THE OCEAN.

BOOK I.—THE SEAS.

	PAGE
CHAPTER I.—General Considerations	1
CHAPTER II.—Oceanic Basins.—Depth of the Seas.—Level of the Surface of the Ocean	5
CHAPTER III.—Composition of Sea-water.—Specific Weight.—Salt Marshes, Natural and Artificial.—Various Substances.—Differences of Saltness.—Marine Salt	22
CHAPTER IV.—Various colours of Sea-water.—Reflections, Transparency, and Proper Colour.—Temperature of the Depths of the Sea	29
CHAPTER V.—Formation of Ice.—Ice-floes, Fields of Ice, and Icebergs.—Ice in the Baltic and the Black Sea	35
CHAPTER VI.—Waves of the Sea.—Regular and Irregular Undulations.—Height of the Waves.—Their Size and Speed.—Ground-swell.—Coast-waves . . .	50

BOOK II.—CURRENTS.

CHAPTER VII.—Great Movements of the Sea.—General Causes of Currents.—The Five Oceanic Rivers	60
CHAPTER VIII.—The Gulf-stream.—Influence of this Current on Climate.—Its Importance to Commerce	64
CHAPTER IX.—Currents of the South Atlantic and the Indian Ocean.—Double Eddy of the Pacific Ocean	78
CHAPTER X.—Lateral Eddies.—Rennell's Current.—Counter-current in the Sea of the Antilles.—Equilibrium of the Waters in the Baltic, the Bosphorus, at the Entrances to the Mediterranean and the Red Sea.—Exchange of Water and Salt between the Seas	85

BOOK III.—THE TIDES.

CHAPTER XI.—Oscillations of the Level of the Seas.—Theory of the Tides . .	92

CONTENTS.

CHAPTER XII.—Theory of Whewell on the Origin and Propagation of Tidal Waves.—Origin of the Tide in each Oceanic Basin.—"Establishment" of Ports.—"Cotidal" Lines 101

CHAPTER XIII.—Apparent Irregularities of the Tides.—Extraordinary size of the Tidal Wave in certain Bays.—Interference of Ebb and Flow.—Diurnal Tides.—Inequalities of Successive Tides 106

CHAPTER XIV.—Tidal Currents.—Races and Whirlpools.—Tidal Eddies.—River Tides 117

CHAPTER XV.—Ebb and Flow in Lakes and Inland Seas.—Currents of the Euripus.—Scylla and Charybdis 124

BOOK IV.—THE SHORES AND ISLANDS.

CHAPTER XVI.—Incessant Modifications of the Coast-line.—The Fjords of Scandinavia and other Countries near the Poles 130

CHAPTER XVII.—Filling up of the Fjords by Marine and Fluvial Alluvium . . 137

CHAPTER XVIII.—Destruction of Cliffs.—The Coasts of the Channel.—The Straits of Dover.—Action of Shingle and Sand.—Giants' Cauldrons.—Spouting Wells on the Coasts.—Tidal Wells 144

CHAPTER XIX.—Undermining of Rocks.—Varied aspect of Cliffs.—Platforms at their Bases.—Resistance of the Coasts.—Breakwaters formed by the Rubbish.—Heligoland.—Destruction of Low Shores 153

CHAPTER XX.—Normal Form of Shores.—Curves of "Greatest Stability."—Formation of New Shores.—Coast Ridges and Sand-banks.—Inland Bays . . 163

CHAPTER XXI.—Shallows of the Coast.—Deposit from Calcareous Rocks.—Appearance of Strands and Beaches 181

CHAPTER XXII.—Origin of Islands.—Islands of Continental Origin.—Rocks of the Shores.—Islands of Depression, Elevation, and Erosion.—Islands of Oceanic Origin.—Atolls and Volcanoes 189

BOOK V.—THE DUNES.

CHAPTER XXIII.—Dunes resulting from the decomposition of Rocks.—Formation of Moving Dunes on the Sea-shore.—Symmetrical disposition of Ridges of Sand 198

CHAPTER XXIV.—Height of the Hillocks.—Advance of the Dunes.—Displacement of "Etangs."—Disappearance of Villages 207

CHAPTER XXV.—Obstacles opposed by Nature to the Progress of Dunes.—Fixation of the Sands by Seeds 215

PART II.—THE ATMOSPHERE AND METEOROLOGY.

BOOK I.—THE AIR AND WINDS

CHAPTER I.—Air the Agent of the Vital Circulation of the Planet.—Phenomena of Reflection and Refraction.—Mirage 220

CONTENTS.

	PAGE
CHAPTER II.—Weight of the Air.—Height of the Upper Strata.—Barometric Measures	228
CHAPTER III.—Mean Pressure of the Atmosphere under various Latitudes.—Density of the Air in the Northern Hemisphere.—Diurnal Oscillations of the Barometrical Column.—Annual Oscillations.—Irregular Variations.—Iso-barometric Lines	232
CHAPTER IV.—General Law of the Circulation of Winds.—Trade-winds from the North-east and South-east.—Equatorial Calms.—Oscillation of the System of Winds	239
CHAPTER V.—Counter Trade-winds or Returning Winds	245
CHAPTER VI.—The Trade-winds of the Continents.—The Monsoons.—Etesian Winds	253
CHAPTER VII.—Land and Sea Breezes.—Winds from the Mountains.—Solar Breezes.—Local Winds.—The Simoon, Scirocco, Fœhn, Tempests, and Mistral	259
CHAPTER VIII.—Zone of Variable Winds.—Struggle of Opposing Winds.—Mean Direction of the Atmospheric Currents.—Law of Gyration	266

BOOK II.—HURRICANES AND WHIRLWINDS.

CHAPTER IX.—Aerial Eddies.—Cyclones of the Equatorial Regions.—The "Great Hurricane"	273
CHAPTER X.—Speed of the Revolving Masses of Air.—Speed of the Cyclone.—Fall of the Barometric Column.—Irregularities of the Wind in the Path of the Cyclone.	280
CHAPTER XI.—Spiral of the Hurricanes in the Two Hemispheres.—Theory of Cyclones.—Nautical Instructions to avoid Hurricanes	287
CHAPTER XII.—Eddies of Tempests.—Whirlwinds	291

LIST OF COLOURED ILLUSTRATIONS.

SECTION I.

	TO FACE PAGE
Straits of Dover	14
Submarine Plateau of the British Isles	15
North Atlantic Ocean	16
Antarctic Land	46
Current of the Gulf-stream	64
Oceanic Currents	100
St Michael's Bay	110
Fjords of Norway	132
Depth of the Zuyder Zee	161
Headlands of North Carolina	176
Dunes of La Teste	208
Rains of Volcanic Ashes	246
Range of the Hurricanes of August and September, 1848	289

THE OCEAN;

THE ATMOSPHERE, METEORS, AND LIFE.

PART I.—THE OCEAN.

BOOK I.—THE SEAS.

CHAPTER I.

GENERAL CONSIDERATIONS.

To the majority of mankind grouped in crowded populations on the continents, extending over scarcely a quarter of the surface of the globe, the sea is little else than a vast abyss without limits or bottom. Even learned men are inclined, by an illusion of intellectual optics, to give a much greater geographical importance to the phenomena of continental regions, than to those of the ocean. Just as our ancestors, beholding infinite space filled with stars and nebulæ arched over their heads, imagined this immensity to be a dome resting on the vast structure of the earth.

Although the influence of the ocean in the general economy of the globe has not been studied with the same care, relatively, as the effect of the rivers which flow through the plains, or of the springs which gush from the clefts of the hills, yet it is still of the first importance, and on it all the phenomena of planetary life depend. "Water is the chief of all!" exclaimed Pindar in the early days of Hellenic civilization; and since then science has revealed to us, that the continents themselves are elaborated in the bosom of the seas, and that without them earth, like a metallic surface, could give birth to no organic life whatever. Thus, as almost all the cosmogonies of primitive nations poetically declare, earth is "The daughter of ocean."

This is not simply a myth, it is a fact. The study of the strata of the earth—rocks, sand, clay, chalk, conglomerates, proves that the materials of the continental masses have in great part been deposited at the bottom of the sea ; and have assumed their form and character there. Many rocks, especially the granites of Scandinavia, which were formerly believed to have emerged in a plastic state from the interior of the earth, are perhaps in reality ancient sedimentary strata slowly transformed by mechanical and chemical action, which operate incessantly in the great laboratory of the globe. Even on the sides and summits of the highest mountains, now raised thousands of feet above the level of the ocean, may sometimes be found traces of the action of the sea in ancient times. Under our very eyes the immense work of creation, commenced by the seas in the earliest epoch, is carried on without relaxation ; with such energy, in fact, that even, during this short life, man may witness important changes along their shores. If the waves undermine and slowly destroy a peninsula here, elsewhere they spread out sandy beaches and form islets. New rocks, differing in arrangement and appearance, succeed the ancient rocks demolished by the waves. Thus the promontories of granite are disintegrated by the action of the waters, which carry away its various constituents, quartz, felspar, and mica, building them up into new rocks. In the same way the clay resulting from the slow decomposition of the porphyritic or granitic felspar is transformed into slate, which becomes sooner or later as hard as the ancient schists. But the dashing waves and the flowing rivers are not the only agents occupied in the formation of new rocks in the bosom of the sea. There is another ever active agent engaged beneath its waters. This agent is animal life. Shells, corals, and innumerable animalculæ with calcareous or silicious coverings, inhabiting the ocean, are incessantly engaged in consuming and reproducing. They absorb and digest matter which the rivers bring down to the sea, and secrete substances which form their skeletons and cases: as, generation after generation, these swarms perish, their remains are spread out over the bottom of the sea or heaped up on the strand ; and at last form immense banks and submarine plateaux which by some subsequent elevation will be brought to light.

Owing to this ceaseless renewal of the rocks, the ocean is constantly creating a world differing from the old one in the appearance and the disposition of its beds. Thus, to the geologist, the invisible depths of the sea should not be of less importance than the exposed surface of the continents. The ground which to-day bears us and our cities, will disappear as the continents of former epochs have already entirely or in part disappeared; and the unknown spaces which the waters now cover, will rise in their turn, and appear as continents, islands, or peninsulas.

In the long period of geological centuries or ages during which the lands are bathed, not by the waters of the sea, but solely by the waves of the atmosphere, the ocean does not the less continue to modify the configuration of the globe by its clouds, its rains, and all the meteoric influences which take their birth at its surface. All those atmospheric agencies which rage about the summits of mountains, riving them and little by little lowering them, it is the sea which despatches them. All those glaciers which polish the rocks, and carry down into the valleys those piled-up boulders, it is the clouds from the ocean which deposit them in the form of snow on the summits of the mountains. All those waters which penetrate by fissures into the depths of the ground, which dissolve the rocks, hollow out the caverns, bring mineral substances to the surface, and cause at times great subterranean subsidences, what are they, but marine vapours returning in a fluid state towards the basin from whence they arose? Finally, the numerous rivers which spread life over all the globe, and without which the continents would be deserts wholly uninhabitable, are nothing else than a system of veins, and veinlets, which carry back to the great reservoir of the ocean, the waters distributed over the soil by the arterial system of clouds and rain. It is, then, to the phenomena of this oceanic life we must attribute the immense geological operations of rivers, and the exceedingly important part which they play in the flora and fauna of different countries, and in the history of humanity itself. The future discoveries of geologists and naturalists will tell us also what share in the production and development of those germs of vegetable and animal life, which reach their greatest beauty on continents, may be referred to the ocean.

As for climate, upon the varieties of which all that lives upon the earth depends,—does it not follow from movements of the ocean, as well as from the position and elevation of the masses of land? The cold of polar latitudes would be more rigorous, and the heat of the tropics more intense, and these extremes would undoubtedly destroy most of the beings now in existence, if the currents of the ocean did not convey water from the poles to the equator, and from the equator to the poles; thus constantly tending to an equalization of temperature. In the same way, the atmosphere of continents would be completely deprived of vapour, and so perhaps rendered unfit for breathing, if the humidity it derives from the sea were not spread by the winds all over the globe. Thus the ocean blends the contrasts of climate, and makes a harmonious whole of all the distinct regions of our planet; it awakens and preserves life on the earth, which it has deposited layer by layer, which it waters by its vapours, and renders fertile by its springs and its rivers.

CHAPTER II.

OCEANIC BASINS.—DEPTH OF THE SEAS.—LEVEL OF THE SURFACE OF THE OCEAN.

The seas which cover the greater part of this planetary sphere have not completely enclosed basins. They all have their origin in the great common reservoir of the Antarctic Ocean, and communicate with each other by wide straits, or by sheets of water of secondary importance. This partial absence of boundaries, and their enormous extent of surface, deprive the seas of that harmony of form observable in the continental masses. Yet, wherever the water washes the shores of the land, it necessarily reproduces its contour; and, in consequence, the sea everywhere presents a distribution exactly the opposite of that of the earth. The twofold basin of the Atlantic with its wide central expansion, corresponds to the two continents of America with their narrow uniting isthmus. The Pacific itself is divided by its immense Archipelago into two vast distinct seas; and the Indian Ocean in the south balances the mass of Asia in the north. Whilst limiting with its waves the shores of the land, the ocean penetrates far into the interior, either by large rounded gulfs like those of Guinea and Bengal, or by seas bordered with chains of islands and islets, like the China Sea and that of the Antilles, or by an intricate network of channels like those of Sunda, and the Polar Archipelago of America. Certain seas also are almost completely enclosed, and communicate with the remainder of the ocean only by narrow outlets, as is the case with the Mediterranean and the Red Sea.

The bottom of all these seas is not always horizontal or even regularly inclined. It is certain that the bed of the sea has, like our continents, but in a far less degree, plateaux, valleys, and plains. Geology reveals to us that in the course of ages the highlands of the continents sink beneath the oceanic expanses, and the abysses formerly hidden by their waters emerge to the light and reveal the inequalities of their surface. Were not the plains and the hills, which now bear our cities and our harvests, in past ages covered by the waters of the deep? Do we not see on the flanks of the Himalayas, 18,000 feet above the level of the mouth of the Ganges, shells which the sea has there deposited in the strata? And do not our navigators search the bottom of the ocean, and, so to speak, investigate its inequalities with those enormous "feelers," their sounding apparatus?

We may well imagine that the submarine surface still preserves all its primitive rudeness; and that its rocks, cliffs, and fells uniformly present edges unworn and sharp, the marks of fracture, just as on the day when the solid rock was first cleft. And, in fact, in the depths of the sea there are no frosts to break off projecting peaks, no lightnings to split, no glaciers to carry them or crumble them away, no meteoric influences to corrode and round them. Nevertheless, if there are not in the sea, as on the land, agencies like these, ceaselessly at work leveling projections, there are others which as ceaselessly labour to smooth the asperities of the surface. There are the sedimentary deposits brought down by the rivers; and innumerable millions of the skeletons of animalculæ, which live in the deep, or fall like snow from the upper strata of the water and gradually fill up the submarine valleys. Those fantastic mountain-chains drawn on the bed of the sea by Buache and other geographers, cannot therefore really exist, since the geological agencies at work under water differ from those which carve out the table-lands and mountains on our continents. If some immense eddy prevented the particles from being deposited in the deep parts of the ocean, then the rocks and the rifts of the abysses would keep their first form, like those peaks and craters of the moon which are not worn away by the inclemencies of an atmosphere. There are, indeed, tracts in the sea where, perhaps from the influence of a submarine counter-current, the rocks of the bottom are not covered by organic alluvium. In the deepest part of that great arm of the sea which separates the Færoe Islands from Great Britain, Wallich drew up from a depth of more than 600 fathoms* a large fragment of quartz detached from the living rock, and several pieces of basalt; it is quite possible, however, that these fragments had been dropped there by an iceberg.

In general, the sea-bed extends for wide spaces in long undulations and gentle slopes. Sailors, who are carried swiftly over the water by wind or steam, and who generally take soundings at places far distant from one another, are tempted to exaggerate the magnitude of inequalities in the sea-bed, and to see chasms and precipices, where the declivity is in reality inconsiderable. Escarpments, similar to those of the continental mountains, very rarely present themselves; Fitzroy was greatly surprised to find in the neighbourhood of the

---

* Marine soundings are always taken in fathoms; a fathom is equivalent to 2 yards or 6 feet linear.

SUDDEN CHANGES OF DEPTH.

Abrolhos, near Brazil, such rapid slopes, that the lead on one side of the ship indicated from 4 to 6 fathoms only, while on the other side it marked from 16 to 22 fathoms. Sometimes a special cause explains these abrupt changes of the level. Thus M. de Villeneuve-Flayose discovered in the Gulf of Cannes, a spring of fresh water springing from the depths of a kind of well, the sides of which sloped at an angle of 27 degrees. But how can we explain that singular gulf which extends immediately in front of Cape Breton on the coast of the Landes? Ought we to attribute its formation to the meeting of the tides, which takes place in the channel of the

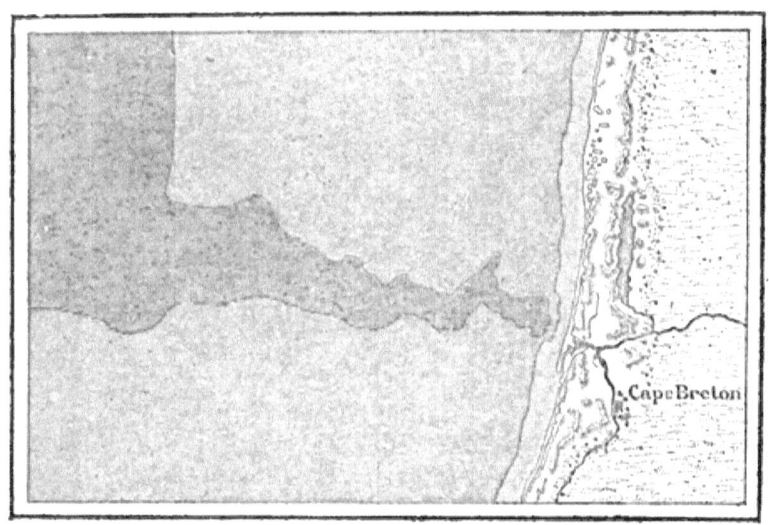

Depths under 3 fath. Depths of more than 64 fath.
Depths from 3 fath. to 54 fath.

Fig. 1.—Gulf of Cape Breton.

Gulf of Gascony? This is a question which it is not yet possible to decide.

We can form some notion of the submarine tracts by surveying the countries that have emerged from under water at a comparatively recent epoch. The Landes of France, the low lands which have replaced the Gulf of Poitou, a great part of the Sahara, the pampas of La Plata, furnish remarkable examples of the regularity of inclination which generally characterizes the bottom of the sea. Even rocky coasts, like those of Scotland and Scandinavia, have been levelled here and there in their lower parts, that were not long ago covered by the waters of the Atlantic. If earthquakes and fissures of the soil, volcanos and slow oscillations of the terrestrial crust, did not on their side increase the inequalities of our planet's surface; it is certain that the incessant contribution of fluvial deposits, the fragments of rocks worn away by the waves, and above all those remains of swarming organisms which fill the sea, would have effected as an inevitable result the equalization of the ocean-beds, and the transformation of their abysses into scarcely indicated slopes; the waters, on their side, would gradually invade the surface of the continents, till, after the operations of myriads of centuries, the earth would become again what it formerly was, a spheroid with its surface entirely covered by a bed of water of uniform thickness.

According to an ancient popular opinion, which, in default of direct observation, was not more contradictory to good sense than many other hypotheses called scientific, the sea was "bottomless;" and this proverbial expression is still that which best conveys to many ignorant persons the real state of things. At the commencement of last century Marsigli himself spoke of "the abyss" of the Mediterranean as of a gulf absolutely unfathomable.* On the other side, mathematicians, supported by theoretical considerations, have attempted to estimate by calculation the average depth of the seas. Buffon, who does not quote the Italian author from whom he has borrowed his argument, gave to the ocean a depth of water equal to 230 toises, or 240 fathoms.† The astronomer Lacaille, whose estimates are no nearer those that recent soundings have rendered probable, allowed from 164 to 273 fathoms of depth to the sea. Laplace, who erroneously estimated the mean elevation of the land at 3280 feet (that is to say, three times the height now approximately determined),‡ thought that the waters of the sea must also be of about equal depth. Young, drawing his deductions from the theory of the tides, assigned about 2735 fathoms to the waters of the Atlantic, and from about 3250 to 3800 fathoms to those of the South Sea. Arnold Guyot remarked that this depth assigned to the Atlantic, would be in fact that of the trench formed in this marine valley, between the coasts of South America and Africa, having the plateaux of Bolivia on the one hand, and those of the Lupata mountains on the

---

* *Histoire de la Mer*, p. 10.   † *Théorie de la Terre : les Fleuves.*
‡ Humboldt. See in Vol. I. the section entitled, *Harmonies and Contrasts.*

other.* This last estimate has, however, only a relative value: if we apply it to the Pacific, continuing westward and eastward the coasts of Asia and America, we should find as the lowest point, and lying (according to this hypothesis) to the east of Easter Island, a depth of

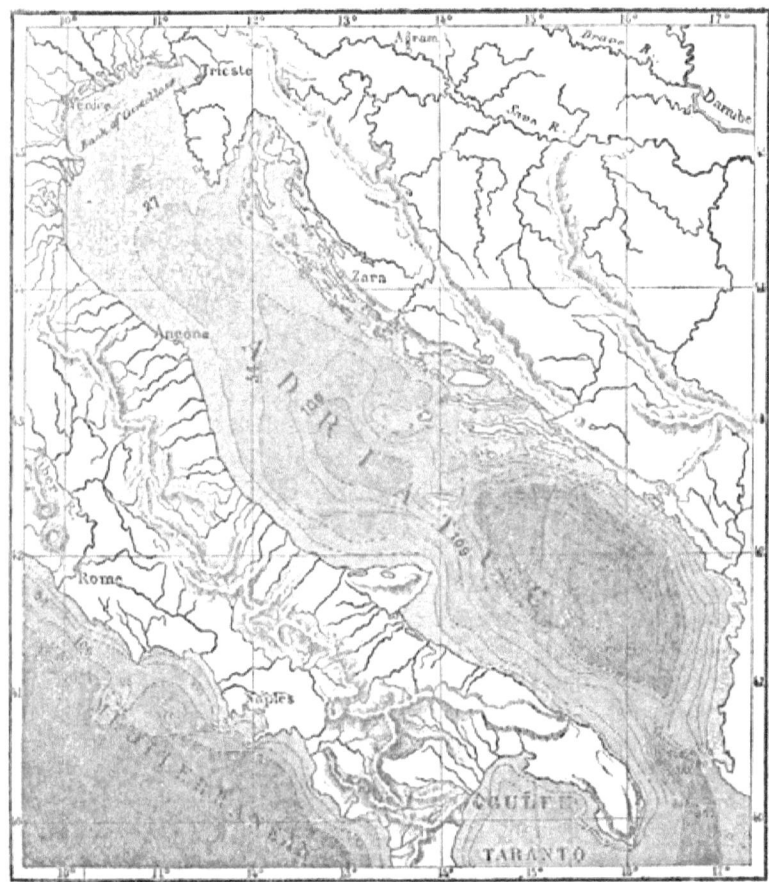

Fig. 2.—Depths of the Adriatic.
(The parts marked by cross-shading are 270 fathoms and upwards in depth.)

about 15¼ miles, three times the elevation of the highest mountain in the world. Evidently it is by direct observation that we must hope, some day, to know all the projections and undulations of the bottom of the ocean; but the instruments which seamen can command are still

* *Earth and Man*, pp. 76, 77.

imperfect, and, except for inconsiderable depths, do not give results of rigorous accuracy. In those latitudes where the water is many hundreds, or even thousands of fathoms in depth, they cannot risk the taking a sounding unless the atmosphere and the waves are in an exceptional state of tranquillity; and even then the slenderness of the cord, the weight of the apparatus, the enormous pressure it endures as it descends, and which increases at the rate of one atmosphere for every 11 yards of immersion, and finally, the long hours which must be employed in this delicate operation, greatly endanger the final success. Unless instruments furnished with electrical bells, like those of Schneider or of Gareis and Becker,* and others more easily employed, more rapid and sure, are used, "bathymetric" observations will be always at great distances from each other, and it will not be possible to construct a submarine map in relief, such as is being constructed of the surface of the continents. Besides, it is very rarely that sailors take soundings in the deep seas simply for the scientific pleasure of investigating the depth of the ocean. It is solely for the requirements of navigation, of commerce, and of industry, that they have observed the depth of the sea, either in gulfs like the Adriatic, or in parts that are filled with sand-banks like the North Sea, in the neighbourhood of coasts and rocks laid down in ancient maps, or in those parts of the ocean which are destined to receive electric cables. In the open sea ships sail almost entirely over unmeasured depths.

Owing to its elongated form and to the amphitheatre of lofty mountains, which all but wholly surround it, the Adriatic offers a very remarkable example of the continuation of the continental slopes below the level of the sea. The bed of the northern part of this gulf, which is a continuation under water of the level plains of Venetia, has an exceedingly gentle slope, much less, in fact, than that of the plains of Lombardy, which seem horizontal.† The sounding-lead shows only a depth of 54 fathoms beyond the narrows formed by the islands of Zara and the headland of Ancona; thus more than a third of the Adriatic is found not to exceed in mean depth rivers like the Mississippi and the Amazons. Farther south the submarine declivity, which continues on one side that of the Apennines, on the other that of the Alps of Dalmatia, becomes comparatively greater, and the sounding-lead descends to about 110 and even 170 fathoms below the surface. At this spot the sea forms a sort of hollow, bounded on

* *Physiographie des Meeres*, 1867.
† G. Collegno, *Geologia dell' Italia*, p. 12.

the south by a submarine isthmus uniting the peninsula of Manfredonia, with the isolated rock of Pelagosa, and with the islands of the Dalmatian coast, Lagosta, Curzola, and Lesina. Beyond this isthmus, and extending as far as the Straits of Otranto, is another and much deeper hollow, towards the middle of which the soundings indicate a depth of nearly 500 fathoms; and on the east rise the precipices of Montenegro, the roots of which descend very rapidly beneath the waters. Thus the soundings of the Adriatic confirm the observations, made long ago by Dampier and many other navigators, that the sea is generally deep at the base of abruptly sloping mountains; and, on the other hand, that there is but a slight depth of water near low coasts.

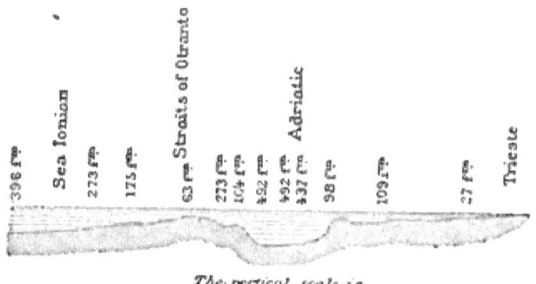

Fig. 3.—Profile of the bed of the Adriatic.

As to the Mediterranean properly so called, it is almost unknown, except in those parts which have been explored for the laying of telegraphic cables; however, on comparing with one another all the soundings, and the various tracks followed by those who have laid the wires, we can at least form a general notion of its submarine surface. If the waters of the Mediterranean were suddenly lowered about 110 fathoms it would be divided into three distinct sheets of water: Italy would be joined to Sicily, Sicily would be united by an isthmus to Africa, the Dardanelles and the Bosphorus would be closed, but the outlet of Gibraltar would remain in free communication with the Atlantic Ocean. If the level was lowered by about 550 fathoms the Ægean, the Euxine, and the Adriatic would wholly disappear, or only leave in their beds unimportant pools; the remainder of the Mediterranean would be divided into several seas like the Caspian, either isolated, or communicating with each other by narrow channels, and the terminal promontory of Europe would be joined by the isthmus of Gibraltar to the mountains of Africa. A depression of about 1100 fathoms would leave nothing but three inland lakes; to the west a triangular basin occupying the centre of the depression between France and Algeria; in the middle, a long cavity extending from Crete to Sicily; and eastward, a hollow lying in front of the Egyptian coast. The greatest depth of the Mediterranean, exceeding 2200 fathoms, lies to the north of the Syrtes almost in the geometrical centre of the basin.*

It is the same with the North Atlantic Ocean as with the Mediterranean. The depth of the central valley, extending from north to south between Europe and the New World, is only known in a general manner. But the gulfs and straits which project from the ocean between the northern countries of Europe, such as the Channel, the North Sea, the Cattegat, the Baltic, have been almost completely explored by the sounding-lead.

The North Sea in all its northern part, from the 51st to the 57th degree of latitude, presents a mean depth of only about 16 to 27 fathoms, except near Newcastle-upon-Tyne, where the bottom is found to be from about 49 to 65 fathoms below the surface. Vast tracts of sand and mud,—the White bank, the Black bank, the Brown bank, the Dogger bank, the Fisher bank, separated from one another by fosses and lateral channels, deeper by from about 6 to 11 fathoms, almost entirely fill its bed, and stretch as far north as the Shetland Islands. There, as in the centre of a whirlpool, is deposited the marine alluvium, whilst that arm of the ocean follows the precipitous shores of Scandinavia over the rocks and compact clays of the bottom. In these parts the lead descends to about 164 and even 437 fathoms

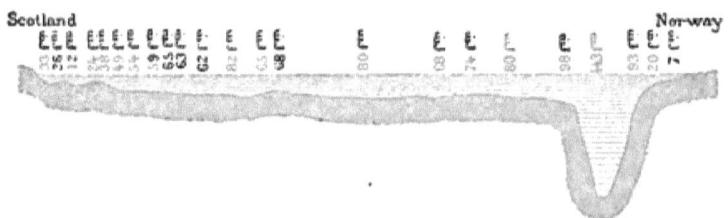

Fig. 4.—Profile of the bed of the North Sea, from the north point of Scotland to Stavanger in Norway.

from the surface of the sea; and in the centre of the Skagerrack, between the sandy beach of Jutland and the bold shores of Norway, nearly 443 fathoms have been reached. One seems to see here,

* Böttger, *das Mittelmeer* ;—*Mittheilungen von Petermann*, 1866.

in vaster proportions, a repetition of those narrow and deep trenches which surround isolated rocks left standing out on flat sandy shores.

From the Skagerrack to the Cattegat, which may be considered as the submarine threshold of the inland waters of the Baltic, the transition is effected somewhat abruptly. The Cattegat presents nowhere more than 93 fathoms, the mean depth of its channel is only 54 fathoms, and the banks of sand and mud render its navigation difficult. The depth of water is reduced to 16, 11, and even in some places to 5 fathoms, in the Sound and the Great Belt, which form the entrances to the Baltic Sea, properly so-called. This vast reservoir, which partakes at the same time of the nature of a Gulf by its free com-

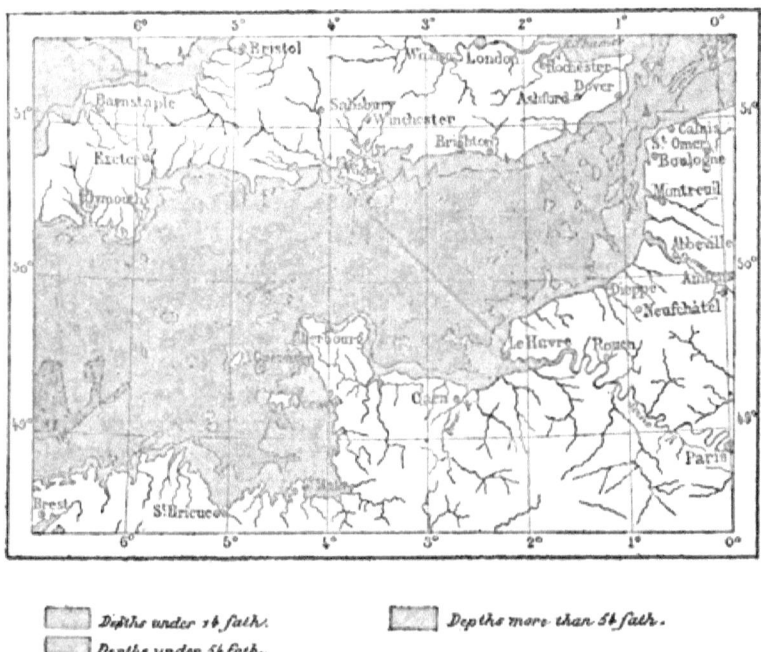

Depths under 16 fath.
Depths under 56 fath.
Depths more than 56 fath.

Fig. 5.—Depths of the English Channel.

munication with the ocean, and of an inland lake by the slight saltness of its waters, has a mean depth of 22 to 33 fathoms, analogous to that of the Cattegat. According to Foss the greatest depth (between the island of Gothland and Esthonia) would be found at only 98 fathoms below the surface of the sea; but according to Anton von

Etzel the lead would not reach the bottom at less than 150 fathoms in the deepest hollow.

To the south-west, the North Sea communicates by the Straits of Dover with the Channel; a narrow arm of the sea which may be considered as a mere accident of the earth's surface, as a kind of maritime trench, so inconsiderable is its depth compared with that of the ocean. In order to form a true notion of the depth of the Channel, compared with its width, one must imagine a miniature of this sea drawn on a scale of one yard for two-thirds of a mile, on a perfectly

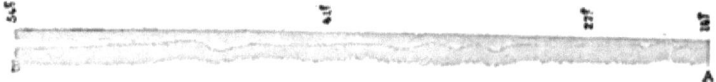

Fig. 6.—Profile following the line of the greatest depth.

horizontal surface. This sheet of water would not have less than 547 yards of length, and its width would vary, according to the coast lines, between 36 and 240 yards. And yet, notwithstanding this considerable surface, the greatest depth would be less than 2 inches at the entrance. In the deepest hollow of the Channel, between the hillock representing Start Point and that of the Sept-Iles, it would be less than 2½ inches. A sparrow could hop this miniature sea.* We see that it is as easy to exaggerate the importance of the depth of the sea as it is the height of mountains. The section annexed to Plate I. shows the proportionate depth of water between the shores of Dover and Calais.

On leaving the Channel, the parts of the oceanic bed which have been explored by sounding are more and more distant towards the west, and then become quite rare. Finally, many hundred miles out at sea, where the true abysses commence, soundings have been only taken at intervals of about 30 and even 55 miles apart. The points thus marked which have served for drawing the submarine chart of the North Atlantic, are therefore by no means numerous, but nevertheless we have in it a pretty exact representation of the *relief* of the ocean-bed. The average depth of water which separates the coasts of North America and those of Europe is about 1915 fathoms, but the central valley presents a surface relatively uniform, and much less varied than that of Europe or even the United States; the greatest slopes do not probably exceed those of the river-beds which seem nearly horizontal; and it may be said that the depth of

* Saxby, *Nautical Magazine*, March, 1864.

The Ocean &c

STRA

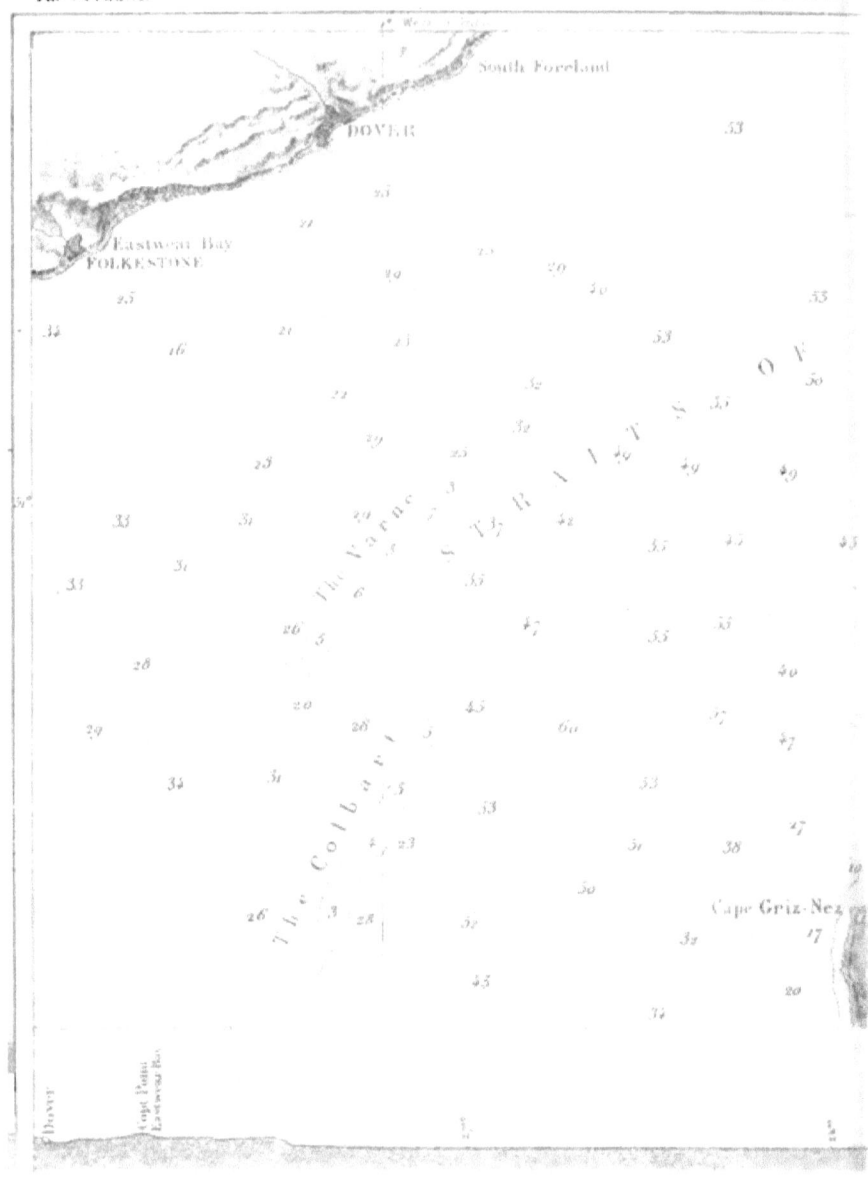

DOVER. Pl. I.

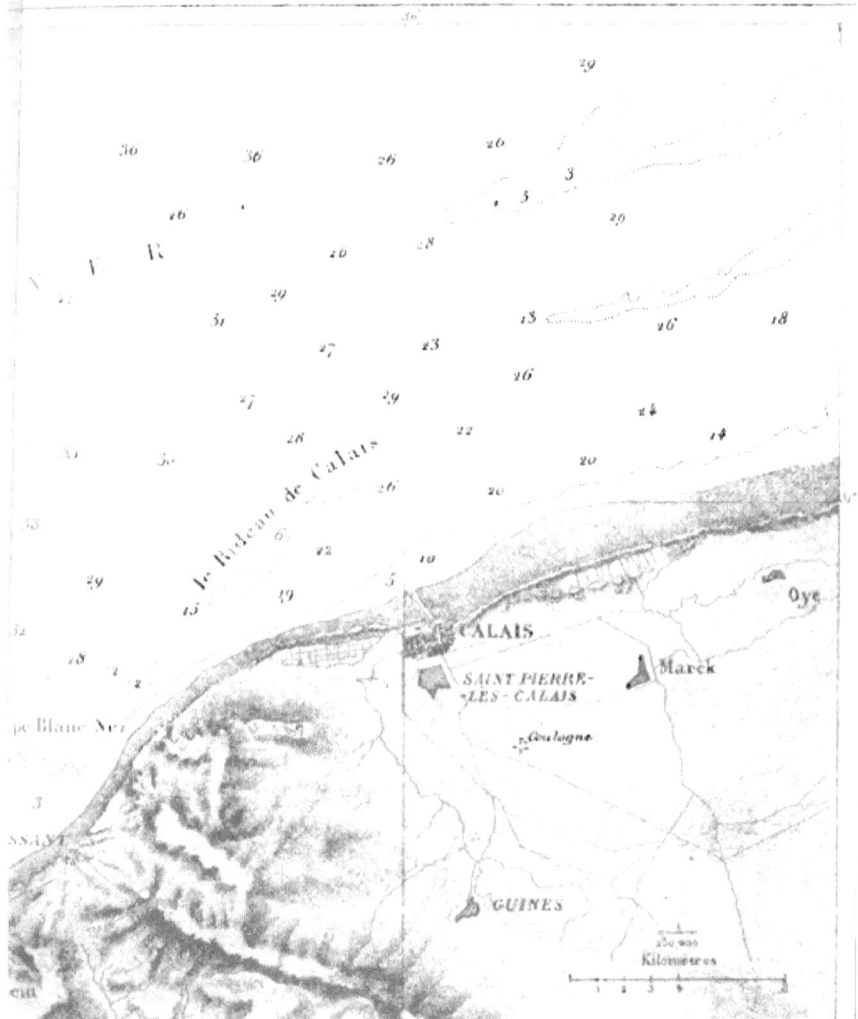

SUBMARINE PLATEAU OF THE BRITISH ISLES.

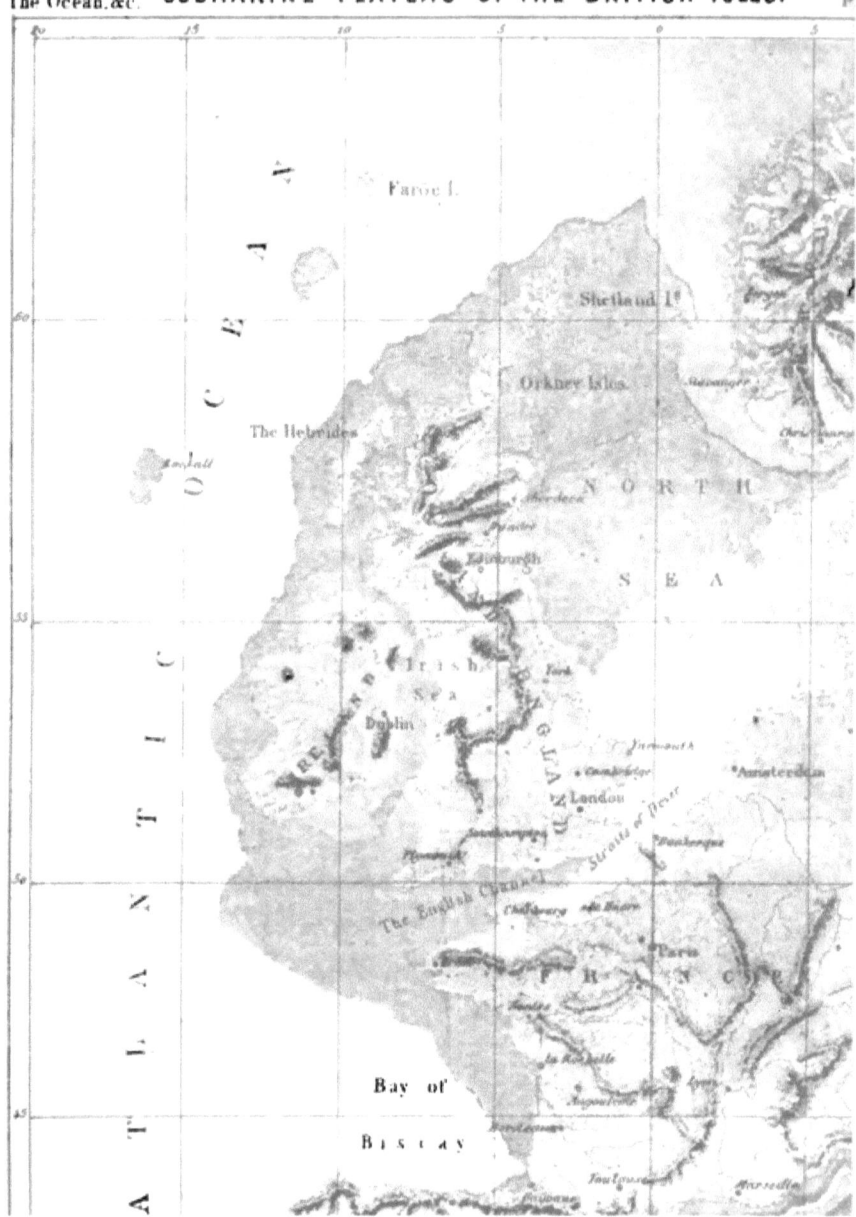

DEPTH OF THE ATLANTIC.

the sea is concentric with the surface. Hence the name of "the telegraph plateau," given by Maury to these plains, some time before the first transatlantic cable was laid. The most considerable depth of this plateau is 4846 yards, that is to say, about a 1639th of the width of this ocean; this being a thickness relatively less than that of the finest needle.* The section on Plate III. enables us to compare the relief of the continental surface and that of the oceanic depths from the coasts of the United States to those of Europe. It is true that in order to render the vertical dimensions visible, it has been necessary to

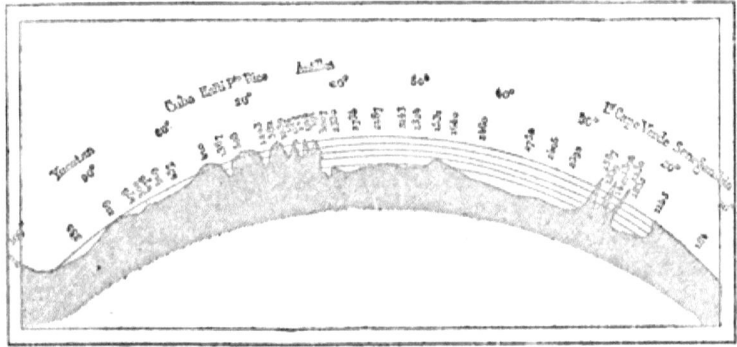

Fig. 7.—Section of the Atlantic in the Tropics.

exaggerate them in the enormous proportion of twenty-fold. To the south the depth of the sea becomes more and more varied. An imaginary section from the plateau of Anahuac to Senegambia across Yucatan, the Caribbean Sea, the Antilles, and the central basin of the tropical Atlantic, presents a much more unequal surface than that of the telegraphic plateau; but the true oceanic part of the basin equally shows a great uniformity in almost its entire extent.

Considered as a whole, the North Atlantic is a depression whose sides descend gradually towards a central hollow situated between the coasts of the United States, the Bermudas, and the bank of Newfoundland. A fall of the waters of less than 110 fathoms would reveal the submarine groundwork upon which France, Spain, and the British Isles rest. This is indeed the true foundation of the European continent, for immediately beyond this basement which forms the extreme angle of the Old World, the bed of the sea, at an inclination of about eight degrees, descends gradually from 110 fathoms to 1610 and 2187 fathoms below the waves. A fall in its level of 1094 fathoms would

* Noto. Bischof, *die Gestalt der Erde und der Meeresfläche,* p. 6, and following.

diminish the width of the Atlantic more than half, would leave the Gulf of Mexico completely dry, and only leave an elongated lake in the central part of the Caribbean Sea. If the present level were lowered by 2187 fathoms, a continent separated from Europe and America by two narrow channels, and extending over a space of from about 1550 to 1860 miles, would stretch into the torrid zone; and, by a remarkable coincidence, would affect that peninsular conformation and southerly direction, presented by Greenland, Scandinavia, Spain, Italy, Greece, Arabia, India, and the three great continents of the South.* A lowering of 3280 fathoms would completely unite Newfoundland to Ireland, and consequently form a bridge between the Old and New Worlds. Even of the central Atlantic there would only remain a narrow "Mediterranean" sea in front of the Antilles and Guiana. Finally, let the waters be lowered by 4375 fathoms, and the northern part of the Atlantic would be reduced to a small triangular "Caspian," situated between the Azores, the bank of Newfoundland, and the Bermudas.

In the present state of science, it is impossible to draw an approximate chart of the depths of the South Atlantic, similar to that which one can construct of the bottom of the northern part. It even appears that many of the soundings made in this part of the ocean must be considered null, because the explorers have not taken into account the deflection of the sounding-line occasioned by submarine currents. The depth of 7600 fathoms obtained by Captain Denham, R.N., is accepted by M. Bischof and other geologists with all confidence, because this explorer took care to raise the cord several times by a hundred yards, and when thrown again it always stopped at the same point. As to the sounding of 8695 fathoms, announced by the American Parker, it is certainly erroneous; since later, in the same latitudes, they have found the bottom at 3007 fathoms only. Not knowing the depth of water in the different parts of the South Atlantic, mathematicians have tried at least to calculate the mean depth of the entire basin by the swiftness of the translation of the tidal waves. They have estimated the mean force of the bed of water in the Atlantic Ocean from the 50th degree south to the 50th degree north latitude to be about 7330 yards. Now, the mean depth of the north basin being very nearly 2187 fathoms, the depth of the southern basin must be estimated, according to this calculation, at about 4920 fathoms.† However, these figures rest on the very

---

* Sir John Herschel, *Physical Geography*, p. 35.
† Sir John Herschel, *Physical Geography*, p. 72.

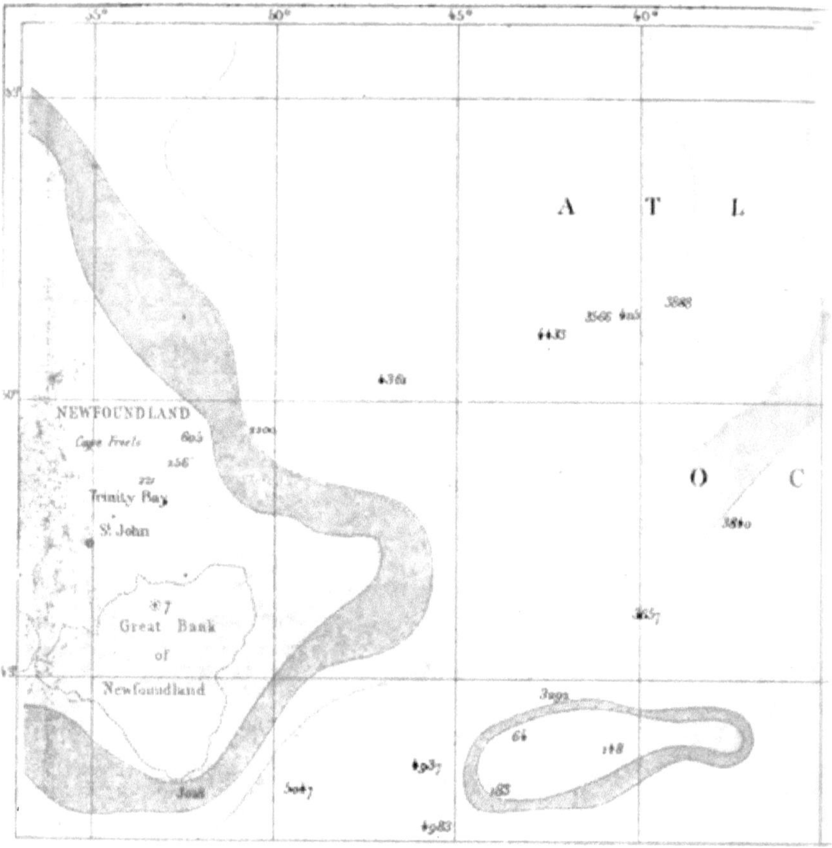

The Ocean. &c.   NORTHERN

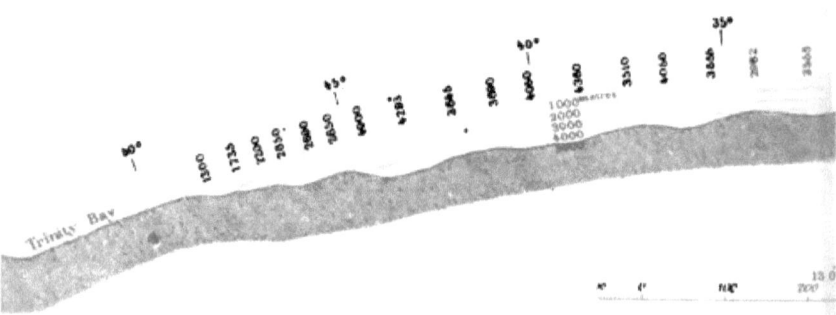

The Scale of heights is in pro

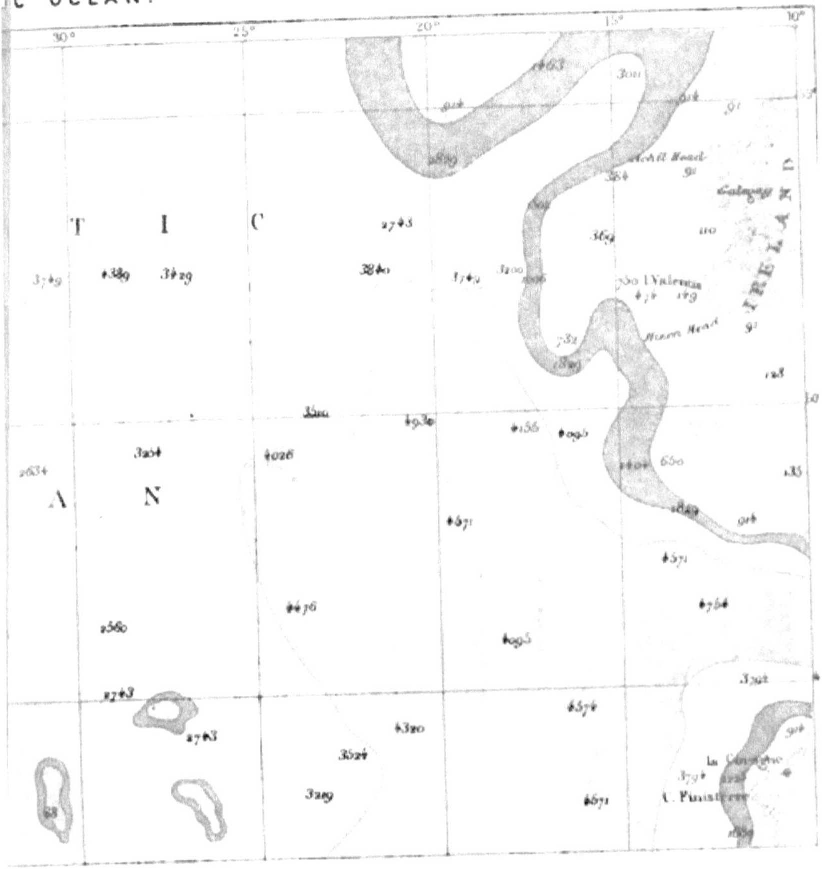

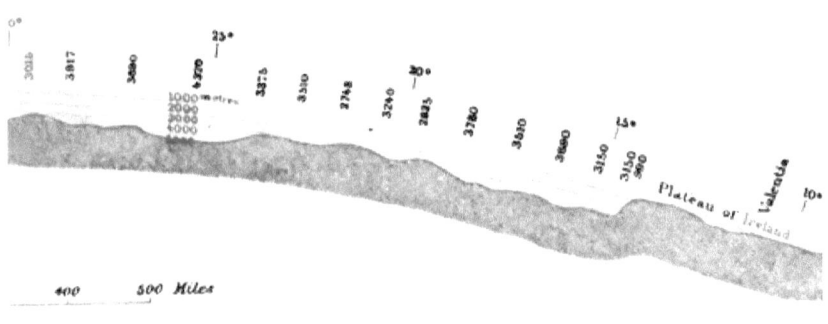

contestable and much contested hypothesis that the tides, instead of forming in a distinct manner in every basin of the ocean, have a common origin in the great South Polar Sea, and roll towards the north like one immense wave, in the double valley of the Atlantic.*

As to that part of the Pacific Ocean comprised between Japan and the coasts of California, it is not by the swiftness of the propagation of the tides, but by that of the earthquake-waves, that the mean depth may be approximately estimated. In the terrible earthquake of December 23rd, 1854, which partially destroyed several Japanese towns, among others Yeddo and Simoda, the vibrations of the marine surface traversed an oceanic space of 6,842 miles in 12 hours and a few minutes; and Prof. Franklin Bache was able to calculate, in consequence, the swiftness of the waves and the depth of the ocean across which they were propagated; this depth is an average of 2,342 fathoms.† Besides, various authentic soundings taken in the northern basin of the Pacific between California and the Sandwich Islands, confirm the result of this calculation, since they indicate a depth varying from 1,968 to 2,570 fathoms. Not far from the coast of California 2,700 fathoms of depth have been found.‡ Between the Philippine and the Marianne Islands two other soundings have given 3,267 and 3,609 fathoms, and even in this last operation the lead has brought up specimens of the submarine soil, and 117 species of minute forms of life. Finally, between the Pacific and the Indian Sea, to the south of the East India Islands, Captain Ringgold found the bottom more than 8¾ miles below the surface. Thus one might throw into this abyss of the sea, not only Pelion on Ossa but Gaourisankar itself, the highest mountain of the globe; and even if on its peak Mont Blanc were set up, the summit of this colossus of the continent of Europe would not reach to the surface of the water.

The Indian Ocean, too, is probably very deep in the greater part of its extent, but we only know those parts nearest land, and those in hardly more than an approximate manner. Its gulfs, like those of the Mediterranean and the Atlantic Ocean, have relatively a slight depth of water: the Persian Gulf, for instance, having a mean depth of only 54, and the Red Sea of 163 to 273 fathoms. Those parts of the Gulf of Bengal which are adjacent to the Coromandel Coast and the delta of the Ganges, increase only very gradually in depth except near the northern extremity of the Gulf, where a prodigious abyss has

---

* See below, the section entitled, *The Tides.*
† *Report of the United States Coast survey.*
‡ Sir John Herschel, *Physical Geography,* p. 39.

been discovered, called "the Great Swatch," which is no less than 2,187 fathoms deep, and is bounded on the north, east, and west by deposits of mud and ooze, which the lead touches at some five or ten fathoms. The formation of this singular funnel is perhaps due to an eddy of tidal waters, commencing precisely at that spot where the alluvium of the Ganges is brought down to mingle with the sea.*

Almost all the Indian Archipelago, Sumatra, Java, Borneo, and the adjacent islands, rest on a submarine bank, having on an average only a depth of 33 fathoms, and even at the deepest places only 55 fathoms. This is probably the base of an ancient continent, of which the innumerable islands, scattered over the sea in these latitudes, are

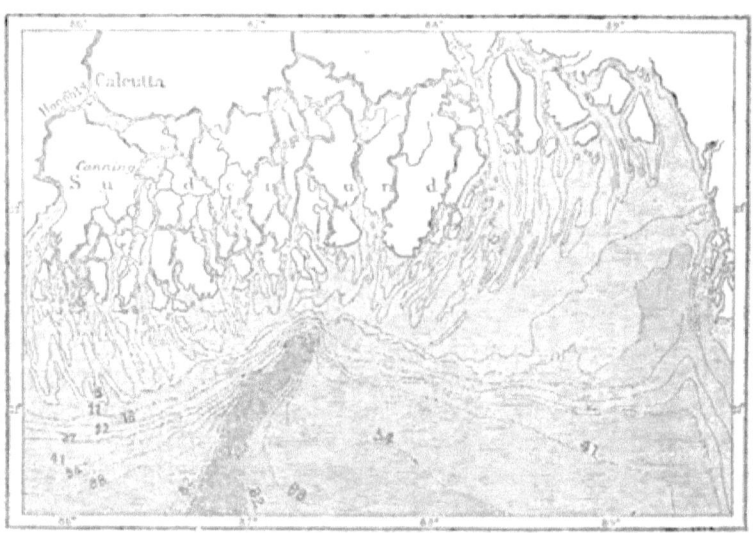

Fig. 8.—Depths of the Sea at the mouth of the Ganges.†

the remains. Another bank, extending for 435 miles to the north and north-west of Australia, supports that continent, and all the neighbouring islands, including New Guinea. A channel of very deep water, not yet sounded, separates from the Asiatic archipelago those higher Australian levels which also seem to be only the ancient fragments of vanished lands.‡ It is around these two great continental basements that the Pacific and the Indian Oceans, properly so called, commence.

* See in Vol. I. the section entitled, *Rivers*.
† The part marked by cross-shading represents the 'Great Swatch.'
‡ See below, the section entitled, *Shores and Islands*, and *The Earth and its Fauna*.

With respect to Antarctic latitudes, 1,722 fathoms have been found between the 63rd and 64th degrees: near the 78th degree, at the very side of the enormous barrier of ice, which hinders any advance towards the pole, Sir James Ross has touched the bottom at 415 fathoms. And this is all the information which navigators have given us. The icy sea of the north is better known, at least in some parts. To the north of Siberia, the bed of the sea, continuing the slope of the hardly-inclined "*tundras*," extends towards the pole with such a slight declivity, that at 156 miles from the coast the lead only shows a mean of from 14 to 15 fathoms. Around Spitzbergen and the western shores of Scandinavia the sea is much deeper, and on the precipitous coast of Norway its abysses join the deep channel which separates Scandinavia from the lesser depths of the North Sea. More to the west, between Scotland and Iceland, the parts explored by McClintock, with the view of laying the telegraphic cable, are rarely more than 328 fathoms, and nowhere present a depth of water of more than 670 fathoms. Between Iceland and Greenland a depth of 1,547 fathoms has been sounded, and in Baffin's Straits are abysses of nearly 2,000 fathoms. This great depression makes Greenland a country quite distinct from the American continent. The plateau upon which this grand island rests presents slopes relatively very steep. On the western side the declivity is in certain places one yard for every five of distance, while the western slopes of the submarine plateau of Ireland, which are among the most rapid in all the ocean, have about one yard of fall for every eight yards of length.*

We can see clearly that the state of our knowledge of the subterranean surface is still very limited; yet the sum of the facts which have been already scientifically confirmed gives a great probability to the opinion, very natural on other grounds, that the oceans deepen gradually towards the south, where the waters occupy the greatest extent on our planet. The celebrated chemist and geologist, Bischof, thinks we may conclude from the comparison of all the soundings, that the bottom of the sea is on an average as near the centre of the globe as the poles themselves. In certain latitudes, and notably towards the 78th degree north, the terrestrial radius drawn to the bottom of the sea is even less than that at the pole, which perhaps is to be attributed to the wearing away of the soil by icebergs. But, on the other hand, in the greater part of the ocean the bottom of the sea is a little more distant from the centre than the poles, which doubt-

* Wallich, *North Atlantic Sea-bed*, p. 18.

less arises from the alluvium brought down by the rivers and the accumulations of organic remains. Thus the part of the globe covered by the seas might be considered as perfectly round, and Newton's hypothesis, explaining the bulging at the equator by the state of fluidity in which the planetary mass had originally been, would become unnecessary.*

As to the mean depth of the whole mass of the marine waters, we can hardly estimate it at less than about 3 miles; since, as we have already seen, the entire basin of the Atlantic, and that of the Northern Pacific, which border upon the great northern continents, are deeper by many hundreds or even thousands of fathoms.

Taking as the total surface of the ocean, an extent of more than 145 millions of square miles, we find that the sea forms a volume of about $2\frac{1}{2}$ million billions of cubic yards, that is to say, the 560th part of the planet itself. Sir John Herschel † gives much higher figures for the same volume of water; but he has taken, as the basis of his calculation, the probable maximum of the depth of the seas, that is to say, four English miles, more than 3,738 fathoms. We cannot speak yet with certainty, but one day, thanks to the new observations which are added every year to those which science already possesses, it will be possible to give figures more relatively exact for the depth of the marine abysses, and the mass of water that fills them. One thing is certain; that the highest part of the continent raised above the surface of the waters is of much less elevation than the depth of the sea; and we can estimate the land above the level of the sea at only about a fortieth part of the mass of waters. Besides which, the land itself contains within it an enormous quantity of water united either chemically or mechanically with all rocks.

The water of the seas urged by the force of gravitation constantly seeks its level, like the water of rivers and lakes. When, in consequence of very rapid evaporation, or of a succession of tempests blowing from the same quarter of the horizon, the surface of the sea is lowered in any gulf, the waters from the adjacent parts rush towards the impoverished space, to fill the void. In the same way, when great rains, the swelling of large rivers, or the action of winds have raised the level of the sea in one point, this local swelling soon subsides, and its superfluity is dispersed over the surrounding surfaces. We may, therefore, consider the mean height of the sea as the same in every ocean, since the natural movement of water tends

---

* *Gestalt der Erde und der Meeresfläche.*   † *Physical Geography*, p. 17.

ever to re-establish an equality of surface in all parts where an accidental disturbance has occurred.

Nevertheless, the diversity of climates, of winds, and of currents, is such, that certain seas, separated from one another by a narrow isthmus, present permanently unequal levels. Thus several German engineers believe that they have established the fact that the Baltic Sea, into which a great number of considerable rivers discharge themselves, is on an average some inches higher (?) than the North Sea.* In the same manner the Atlantic, whose waters spread out on one side into the North Sea, and on the other into the Mediterranean, would have a mean level scarcely higher than that of the two basins which it supplies; while the Black Sea and the Gulf of Venice, receiving, like the Baltic, several large rivers, would, like the latter, be proportionably elevated. On the two sides of the Isthmus of Suez the waters are also found at slightly unequal heights. According to the engineer Bourdaloue, the mean level of the Red Sea at Suez exceeds by $31\frac{1}{2}$ inches that of the Mediterranean near Port Saïd; at low tides the two sheets of water are perceptibly of the same level, while at high-water the sea is sometimes higher by $3\frac{1}{4}$ feet in the Bay of Suez than at the northern extremity of the maritime canal. A similar difference, too, occurs between the Bay of Colon and the Gulf of Panama, and there also it is the mass of water in which the tides have the fullest swell, that is to say, the Pacific Ocean, which has the highest level. But the measurements made on the always unstable level of the sea are very delicate operations, as one can so easily make a mistake at starting, through the oscillations of the ebb and flow; and over spaces of many miles, divided by various obstacles, it is very difficult to avoid slight errors. At all events, it is certain that the surface of the sea, unceasingly traversed and perturbed by winds, currents, and tides, is not perfectly horizontal at any point of the globe.

* Woltmann;—Von Hoff, *Veränderungen der Erdoberfläche*, t. iii. p. 328.

CHAPTER III.

COMPOSITION OF SEA-WATER.—SPECIFIC WEIGHT.—SALT MARSHES, NATURAL AND ARTIFICIAL.—VARIOUS SUBSTANCES.—DIFFERENCES OF SALTNESS.—MARINE SALT.

BESIDES the ooze, the remains of animalculæ, and innumerable fragments held in suspension, the sea-water is also charged with chemical substances in solution, which give it a specific gravity considerably superior to that of fresh water. This varies in all seas, according to the quantity of the substances dissolved, the amount of evaporation, the contributions of rain and rivers, the direction of the currents and counter-currents. In the polar seas the specific gravity of the waters is also modified by the formation, or melting, of the ice. Every variation of temperature, every local movement of the sea, causes a more or less perceptible modification in the proportion of the salts dissolved, and in the specific gravity of the water. Thus we can only obtain an average for the various conditions of the fluid mass in the different seas.

The mean specific gravity of oceans with deep basins is nearly 1028; that is to say, sea-water weighs 2·8 per cent. more than the same bulk of distilled water. In the Mediterranean, where the heat of the sun evaporates more water than the rivers bring down to it, the average specific gravity exceeds 1029; in the Black Sea, on the other hand, where very considerable rivers of fresh water discharge themselves, the specific gravity is reduced to 1016. And all the intermediate degrees between these extreme specific gravities are found, according to the varied physical conditions which exist, in other seas. Furthermore, it seems to be established that the waters of the ocean in the southern hemisphere are, on an average, lighter than those of the northern hemisphere.[*]

The average quantity of all the salts contained in the sea, or the saltness of sea-water, was estimated by Bibra and Bischof at 35·27 parts in 1000; but much more complete observations made since by

---

[*] Horner; J. Davy.—Bischof, *Lehrbuch der chemischen Geologie.*

SALTNESS OF THE SEA.

Forchhammer have reduced this proportion to 34·40. Besides, almost all the analyses, which up to the present time have been made of sea-water, confirm the general opinion of chemists, that the relative proportion of the matters dissolved is the same in all seas. The quantity of common salt (chloride of sodium) dissolved in seawater, is always a little more than three quarters (75 786) of the total mineral matter held in solution.

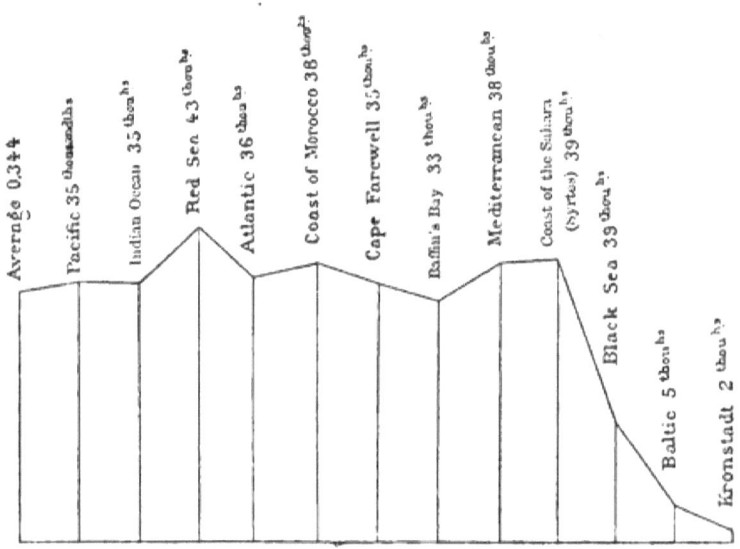

Fig. 9.—Comparative saltness of Seas.

In the north tropical Atlantic, on the coasts of the Sahara and of Morocco, where the sea receives no tributaries, and where, on the other hand, the evaporation is very rapid, the average of oceanic salts is nearly 38 parts in 1000. In mid-ocean, and more especially in the neighbourhood of America, where the water of many great rivers mingles with that of the sea, the saltness is less by one, two, and even three thousandths; but it is generally greater in the tepid waters of the great current called the Gulf-stream, which crosses the Atlantic obliquely. The proportion of salts contained in this current always exceeds 35 thousandths,[*] while the water that flows from the pole towards the equator by Baffin's Bay, contains only about 33 thousandths. It is to the enormous accumulation of ice, that these currents owe the slightly less saltness of their waters. The quantity of

[*] See below, the chapter headed, *Currents*.

cold water which flows from the Antarctic Pole towards the south of Africa and America, contains likewise less saline matter than the seas of the temperate and equatorial zones.

With regard to basins almost enclosed, like the Mediterranean, the Caribbean Sea, and the Baltic, the saltness ought evidently to be greater or less there than in the ocean, according as the evaporation is in excess of or is inferior to the fresh water brought by the rivers and the clouds. In the Mediterranean, the loss in evaporation being more considerable than the contributions of fresh water, the saltness ought to increase in consequence, and the liquid mass would constantly diminish, if a current setting in from the Atlantic through the Straits of Gibraltar did not restore the equilibrium. While the less saline waters of the ocean thus penetrate into the Mediterranean flowing along its surface, a submarine counter-current, composed of heavier and salter water, flows deep below in an opposite direction, and mingles with the waters of the Atlantic, which contain less salt. The mean saltness of the Mediterranean is nearly 38 thousandths, and even exceeds 39 thousandths on the coasts of Tripoli, where the parching winds of the Libyan desert blow.

In like manner the Caribbean Sea seems to present a somewhat high relative saltness because of an excess of evaporation over the contribution of fresh water; but the contrary happens in the Gulf of St. Lawrence, in the North Sea, and above all in the Baltic and the Euxine. The saltness of the North Sea is in different parts from 30 to 35 thousandths, while that of the Baltic, a shallow sea into which so many rivers flow, and where the least wind disturbs the waters,[*] does not quite amount to five thousandths; in the port of Cronstadt it is not even two-thirds of a thousandth, which is almost fresh water. As to the Black Sea, it preserves, even more than the Baltic, the character of a gulf of the ocean, for the average saltness is about half that of the Atlantic.

These differences of salinity between the central basin of the Atlantic and its tributary seas are not in themselves astonishing; but we do not yet know why the South Sea and Indian Ocean contain less saline matter in their waters than the Atlantic, unless the enormous quantity of Antarctic ice explains this difference. While the latter has a saltness of about 36 thousandths, the water of the Pacific has less by nearly one thousandth, and the Indian Ocean contains no more than 35 thousandths of chemical substances. The Atlantic, however, receives a greater quantity of fresh water than the other

[*] Von Sass, *Zeitschrift für die Erdkunde.*

oceans, and the evaporation is probably not so great on an average as in the Indian Ocean. But nevertheless the Gulfs of the Indian Ocean present phenomena analogous to those of the inland seas of the Atlantic. Thus the Red Sea, into which no single permanent stream of water flows, and where evaporation proceeds with very great intensity, shows the enormous degree of saltness of 43 thousandths; such a proportion as is only found in inland salt-water lakes.*

Chloride of sodium or common salt contributes, as we have said, three quarters of the saltness of sea water: this is indeed the characteristic salt of the ocean which most of all gives it its peculiar flavour, and that odour with which the sea-breezes, laden with the fine spray of the waves, are charged. The air which rests on the sea also contains salt to a considerable height; at an elevation of 2,000 feet above the coast on the sides of the mountain which towers above the Peruvian town of Iquique, Mr. Bollaert asserts that any materials washed in distilled water are covered in a few days by a slight incrustation of salt.†

The thickness which a layer of chloride of sodium in the open sea would form if crystallized, would be on an average nearly two inches to every fathom of water, so that if one could imagine the entire evaporation of the waters of the ocean, estimating them to be on the average above three miles deep, there would remain at the bottom of its bed a layer of salt of about 230 feet in mean thickness, which would represent for the whole extent of the seas more than a thousand millions of cubic miles. We can understand how, with such vast quantities of chloride of sodium in solution, the sea has been sufficient to form those enormous beds of rock-salt that are found in the earth in various parts of Europe, without reckoning many other deposits which still remain to be discovered, and which, sooner or later, will be revealed to us by the labours of miners, or by Artesian borings.

Then, too, we may see the ocean at work on all the low coasts, where it deposits saline beds, destined to become in process of time masses of rock-salt, after they shall have been buried beneath more modern strata. When, in consequence of a tempest, or of a high tide, the waters of the ocean are spread in a thin sheet over a flat shore, or in some deeper depression, this slight bed of salt water, spread over a vast surface, evaporates rapidly under the rays of the sun, and leaves in its place a slight white crust of saline crystals. Other

* Forchhammer, *Philosophical Transactions*, part i. 1865.
† *Antiquities*, p. 258.

sheets of water, urged by the billows or the tide into the same basin of evaporation, disappear likewise, forming new layers of crystals; it is thus that real banks of a considerable thickness are gradually formed on the borders of the sea, as well as on the shores of inland seas and salt lakes.*

Even the Black Sea, where the proportion of salt is relatively very

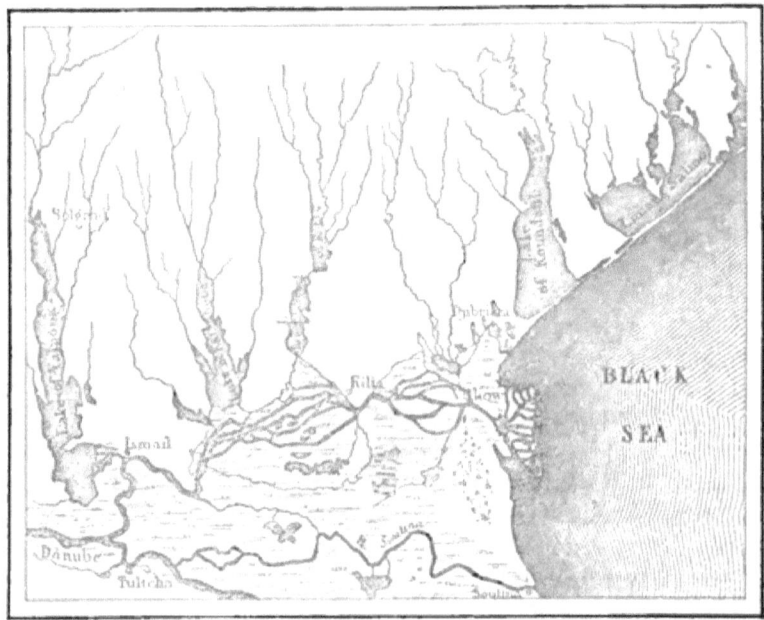

Fig. 10.—Salt Marshes of Bessarabia.

inconsiderable, is, on the greater part of its shores, bordered with these natural salt marshes. In Bessarabia, to the south of Odessa, three "limans" of a total area of many square miles, cease in summer to receive their affluents of fresh water, and all the water which has been brought there in winter, evaporates, leaving an incrustation of salt; towards the centre of the basins of crystallization the solid mass attains nearly an inch in thickness. In 1826, these natural deposits, worked by the natives, produced about 120 thousand tons of pure salt.† In most of the populous countries of Western Europe, man has converted these casual

* See in Vol. I. the section entitled, *Lakes.*
† Bischof, *Lehrbuch der chemischen und physikalischen Geologie.*

swamps into salt marshes with regular outlines. The unequal depressions, where the water of the sea evaporated accidentally, are transformed into reservoirs, where the water is conducted from compartment to compartment, to saturate itself gradually and deposit the pure salt in equal layers. But these are only economical works; man is confined to regulating the operations of the sea itself.*

Besides common salt, many substances which are exceptionally found in inland waters and hot springs, form a part of the normal composition of sea-water. The various simple substances which science has been able to discover therein (either directly by the analysis of the liquid, or indirectly by the study of the plants which draw all their nourishment from the ocean) are twenty-eight in number; but doubtless numerous other simple substances are likewise contained in sea-water, many of which will not long escape the piercing researches of chemists.

After oxygen and hydrogen, which constitute the liquid mass itself, the principal elements contained in sea-water are: chlorine, nitrogen, carbon, bromine, iodine, fluorine, sulphur, phosphorus, silicon, sodium, potassium, boron (?), aluminium, magnesium, calcium, strontium, barium. The common fucus and other sea-weeds contain the greater part of these substances, as well as several metals. They have discovered copper, lead, and zinc, in the ashes of *Fucus vesiculosus*; cobalt, nickel, and manganese in those of the *Zostera marina*. Iron may be obtained directly by an analysis of sea-water, and finally silver is found in a zoophyte, the *Pocillopora*. Forchhammer has obtained from a branch of this coral about a three millionth of silver, mixed with six times the same quantity of copper, and eight times of lead. A slight proportion of silver is precipitated on the bottoms of ships, in consequence of the magnetic current established between the copper-sheathing and the water of the surrounding sea.† And lastly, arsenic has been found in the boilers of steamers which have been supplied with sea-water.‡ It is true that these various substances only exist in infinitesimal proportions in the water, and it is by indirect means alone that chemistry succeeds in revealing them; but the total mass of silver contained in the ocean is estimated at two millions of tons.

The seas having most probably received from the terrestrial strata, which have been unceasingly worn away by the currents of water, all

---

* See below, the section entitled, *The Works of Man*.
† *Philosophical Transactions*, part I. 1865.
‡ Bischof, *Lehrbuch der chemischen und physikalischen Geologie*.

the substances which they contain in solution, we may conclude that the proportions of these substances have continually varied during the geological eras. The saltness would be modified from age to age, according to the various quantities of soluble substances which the rivers brought down to the ocean; and which it returned again to the land, either directly, by depositing them on the shore, or indirectly, by fixing them in the tissues of its plants, corals, and other organisms which people its expanses. By ingenious comparisons between the conditions of the present day and those which seem to have existed in former times in the sedimentary beds, several geologists have attempted to determine if the substances in solution in sea-water have augmented or diminished. But the conclusions at which they have arrived rest, at present, on data too hypothetical for us to regard them as a new conquest of science. It is only certain that in our day the proportions of the substances dissolved have not ceased to vary in every sea. We can judge of this by the enormous difference that there is between the saltness of the waters of the Caspian and those of the Black Sea, two separate basins which formed a part of the same ocean at a geological epoch still comparatively recent.

Sea-water contains also a great quantity of the atmospheric gases, the proportions of which constantly change with heat, light, the motion of the waves, and barometric pressure. Salt water retains dissolved air better than fresh water, and the bulk which it absorbs is generally greater by a third than that found in rivers. It varies from a fifth to a thirtieth, and gradually increases from the surface to a depth of about 325 to 380 fathoms.* Carbonic acid gas is also contained in a relatively very considerable proportion in sea-water, as might have been expected from the swarming myriads of marine animals. Under the influence of light, plants decompose this gas, which diminishes during the day and is increased again during the night. As to the quantity of dissolved oxygen, it varies inversely; during the day it increases by degrees, to be again reduced in the hours of darkness. As by a sort of respiration the great sea, that immensity alive with organisms, absorbs and disengages alternately the gases necessary to the maintenance of life, and measures each breath by the daily course of the sun.

* Bischof, *Lehrbuch der chemischen und physikalischen Geologie.*

CHAPTER IV.

VARIOUS COLOURS OF SEA WATER.—REFLECTIONS, TRANSPARENCY, AND PROPER COLOUR.—TEMPERATURE OF THE DEPTHS OF THE SEA.

OWING to the double property which water possesses of reflecting light and allowing its rays to penetrate to a great depth, it presents successively the most varied colours, the most delicate tints, with alternations the most fugitive and changeable that are to be found in nature. The sea produces and at the same time modifies the varied face of the heavens with all the play and gradation of light and shade. At dawn, the surface of the water is gently brightened by the glimmering of the atmosphere as yet pale and faint; then the sparkling of the waves becomes more brilliant, and the full light of day pours a flood of fire upon the billows. The least movement in the air is betrayed by a change in the aspect of the water, every cloud in passing mirrors itself with the forms and shades of its vapours, every breath of wind that just curls the waves renews the harmony of the changeable colouring on the face of the ocean. And when evening comes, the sea reflects back to the sky all its splendour of purple and flame. It is then that we see on the horizon, "two suns appear one in front of the other."

But the water does not owe its beauty to the splendour of the sky alone, it is beautiful also from its transparency; whilst the substances suspended in the liquid mass, which are visible to a considerable depth, modify by their own colour the general tint of the sea. The animals, fish or cetaceans, which come to the surface or glide swiftly through the waves, cause them suddenly to glitter with changing reflections of grey, rose, green, and silver. The fuci, too, growing beneath the water, vary the aspect of the liquid strata which cover them, and where these beds of plants alternate with ridges of bare rock, or tracts of sand, the sea presents a wonderful mixture of different shades with blended and tremulous outlines. In those latitudes where the water is very transparent, the colour of the ground may be thus distinctly seen at 10, 20, or even 25 fathoms below the surface, which navigators have confirmed by scientific observations made with the greatest care.* But this transparency does not seem to depend upon the intensity of light received, for in the Arctic Seas floating objects can be perceived at as great depths as in the Caribbean Sea; and it is indeed in polar latitudes that the eye of man has been able to penetrate to the greatest depth below the surface. According to Scoresby, that conscientious explorer of the polar seas, the sea-bed of the pure waters in these regions is sometimes visible at a depth of 70 fathoms.† It is true that, in consequence of climatic differences and the organic life which depends on them, the bottom of the sea is much more curious to contemplate in the tropical zone than in the neighbourhood of the poles. There is nothing more delightful than to sail over one of those seas where, without fear of hidden rocks, one can watch the bed of the sea reveal itself far below the prow of the vessel. Numerous algae, green or rose coloured, wave gracefully below the surface like the grasses of a brook; the molluscs crawl along the bottom; crustaceans, fish, star fishes of brilliant colours, and many other animals of strange form, glide slowly or dart like arrows through the blue water, glistening in a thousand changing hues; while the Nemertida and other living ribbons softly unroll their transparent rings. One might fancy oneself suspended above another earth, and floating in an aërial ship. The white foam on the waves raised by the keel of the ship, and the iridescent colours which sparkle in the spray, add fresh charms to this wonderful picture.

Even when the bed of the sea is not distinctly visible, it does not fail to reveal itself by the peculiar tint it imparts to the water. In general the colour of the sea is lighter near the coasts, and even at a depth of above 100 to 150 fathoms, a paler shade of the water at times makes known to the practised eye of the mariner the relative proximity of the bottom. Not far from the coasts of Peru, de Tessan perceived that the sea had suddenly assumed a tint of dark olive-green, and when he caused a sounding to be taken, it was found that the mud at the bottom was precisely of this colour. Numerous navigators have affirmed that in one part of the Lagullas bank, where the mass of water is above 100 fathoms deep, the water passes suddenly from blue to a greenish colour.‡ Lastly, off Loango, the water is always brown, similar to that of the bottom, which Tuckey has found to be of an intense red. Is, then, this colouring owing to

---

* Cialdi, *Sul moto ondoso del mare*, p. 284.
† *Arctic Regions.* See also the Notice of Arago, *Œuvres complètes*, t. ix.
‡ Arago, ibid.

the sun's light, which descends through the liquid depths to the bed of the sea, and is reflected again to the surface; or does it result, as Cialdi thinks, from particles of mud that are floating in the water?*

Another question, difficult to solve, is that of knowing what is the natural colour of the sea-water. Not to mention local colouring, resulting like phosphorescence from numberless minute living organisms,† the various parts of the ocean almost always present, whatever may be the state of the atmosphere, a normal tint easy to be distinguished from accidental shades. Thus, to cite one of the most striking contrasts, the water of the Gulf of Gascony is of a sombre green, while in the Gulf of Lyons the water of the Mediterranean is of a magnificent azure, deeper than that of the sky. The wonderful blue colour which rises from the depths of the water in the grotto of Capri, so frequently visited by travellers, is a well-known example of the degree of intensity to which the blue peculiar to the waters of the Mediterranean can attain. In the tropical latitudes of the Atlantic and the South Sea, the azure of the ocean is no less beautiful than that of the Tyrrhenian Sea; while in the direction of the poles the water gradually assumes a greenish tint. Naturalists have concluded from this fact that the refraction of the rays of light, which are much more vivid under tropical latitudes, play a principal part in the blue colouring of the sea. Maury thinks that the saltness is also one of the causes which contributes the most to give its azure tint to the water; and observes that the Gulf-stream of the American coasts, superior in salinity and in temperature to the water around it, is also of a much deeper blue. In the same way the shallow water let into the salt marshes of coasts, gains in intensity of colour in proportion as the salt is concentrated there. Still it is very possible that the colouring of the sea is due in great part, like the marvellous tints of the Swiss lakes,‡ to innumerable corpuscules held in suspension, upon which the light strikes.

The law of the distribution of temperature, in the depth of the ocean, is not as yet more determined than that of the colouring of the water. At the surface of the sea it is as easy to make observations as in the air, and it has been determined, without difficulty, that this superficial sheet of water presents on an average, in all climates, the same degree of heat as the superincumbent atmosphere. Thus,

---

* *Sul moto ondoso del mare*, p. 287.
† See below, the section entitled, *The Earth and its Fauna*.
‡ See in Vol. I. the section entitled, *Lakes*.

from the polar regions to the equatorial zone, the water becomes warmer with an almost regular gradation, and, from the freezing-point under the Arctic circle, the temperature rises to 68° and 77° Fahr. under the tropics, and to 86° and even above 90° Fahr. in the Pacific, the Red Sea, and the Indian Ocean.* With regard to the increase or decrease of heat in a vertical direction, we had till recently only the vaguest notions, in consequence of the want of exact soundings. It is in fact very difficult to lower to a depth of several hundred, and even several thousand fathoms, thermometrical apparatus strong enough to resist the enormous pressure of one atmosphere for every 33 feet.

Sir James Ross was one of the first who attempted to apply the resources of modern science to a systematic exploration of the temperature in the depths of the sea; but he seems to have committed the error of generalizing too hastily from the incomplete results which he had obtained; and in his eagerness, he believed he had discovered a law, which the subsequent researches of navigators have not confirmed. He thought that he could establish the fact that under the equator the temperature of the water diminishes gradually to 1200 fathoms, where it is only 39·2 degrees Fahr. On each side of the equator the upper waters gradually cool, and the limit of four degrees is progressively raised towards the surface; it is at the fiftieth degree of latitude, in the southern hemi-

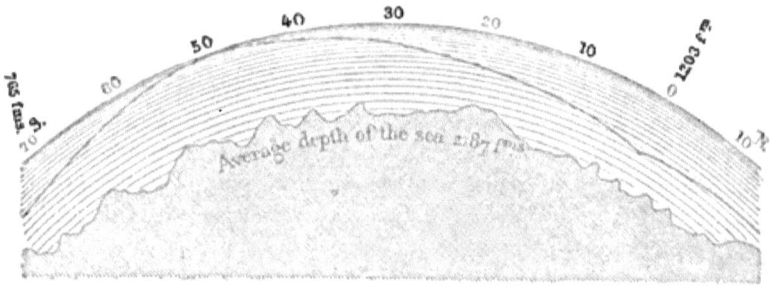

Fig. 11. Sheet of Water presumed to be at a temperature of 39·2 Fahr.

sphere, that it finally reaches the level of the sea. Farther in the direction of the pole the superficial water continues to grow colder, while the line of four degrees sinks gradually to the depth of 765 fathoms. Thus, as the accompanying figure shows, the line of

* Fitzroy, *Weather-Book*, p. 84.

uniform temperature to the south of the equator describes a long curve, touching the surface of the water at one point only. Admitting with the naturalists of his time, that the sea-water has its greatest density, and in consequence its greatest relative weight, at seven degrees above freezing-point, Sir James Ross concluded from this that all the deep waters below this line of 39·2 degrees have the same temperature, and are collected by reason of their condensation at the bottom of the oceanic basins.

Nevertheless it has since been proved, by the observations of Neumann* and other scientific men, that if the greatest density of fresh water corresponds in reality to 39·2 degrees Fahr., the water of the sea only attains this maximum at nearly four degrees below freezing-point (28·4 Fahr.), or even at still lower temperatures, and, in consequence, the conclusions at which Sir James Ross arrived are negatived. Experiments made in chemical laboratories, however, where substances are treated in small quantities, cannot give a perfectly exact notion of the phenomena which have nature itself for their theatre, and which take place either in the aërial spaces or in the vast oceanic basins. Thus, as the celebrated meteorologist Mühry says, the immense sea, and a bucket-full of salt water, do not obey absolutely the same laws of temperature and density. But before the difference is established, nothing can authorize us in maintaining a superannuated theory against all the experiments of chemists, according to which the volume of salt water in the sea in cooling presents phenomena identical with those of fresh-water lakes. Moreover, during the past years, numerous observers of polar seas have found at great depths, beds of water, at a temperature lower than 39·5 degrees Fahr.†

That which remains of the researches of the eminent navigator, Sir James Ross, is that in the tropical and temperate seas the heat diminishes gradually and constantly to a considerable depth. This is what has been put beyond all question, by soundings taken by Fitzroy and other marine explorers. To the south of the island of Madagascar, the surface of the water having then a temperature of 75·2 degrees Fahr., Fitzroy ascertained that the thermometer fell in the most regular manner, till at the depth of 420 fathoms, where they ceased sounding, the temperature indicated hardly exceeded 51·8 degrees Fahr.‡

* *Ueber die Dichtigkeit beim Meerwasser.*
† Fitzroy, *Weather-Book*, p. 81.
‡ *Adventure and Beagle*, Vol. II., Appendix, p. 303.

In the enclosed basins of inland seas thermometrical observations are much more easily made than in the middle of the great ocean; because the waters there are generally less deep, and the natural gradations of temperature are less disturbed by currents. Thus the water is not very cold in the depths of the Mediterranean, and presents only slight variations of temperature. At about 100 to 275 fathoms below the surface, the fluid mass has already attained permanently the mean temperature, which it preserves during all the year, and which seems to correspond to the mean annual temperature of the neighbouring lands, which are subject to all the abrupt changes of heat and cold.* While in summer the superficial sheet of water has about 73·4 degrees Fahr., the water comprised between 273 fathoms depth and the very bottom of the Mediterranean is found at 59 degrees Fahr., which is pretty nearly the mean annual warmth of the bordering countries. In the Greek Archipelago, the deep waters of which are probably colder in consequence of the current flowing from the Black Sea, the waters of the surface have in summer from 77 to 78·8 degrees Fahr., and at hardly 98 fathoms depth the thermometer reveals a constant temperature of from 53·6 to 55·4 degrees Fahr. The Mediterranean is divided into distinct basins, separated from one another by intermediate ridges, which are situated from 98 to 273 fathoms below the surface, the result being that the variations of temperature produced by the movements of currents and counter-currents are arrested on the tops of the ridges. The water of each basin, being relatively tranquil, thus maintains almost constantly the same thermometrical degree.†

* See below, the section entitled, *Climates*.
† Spratt, *Nautical Magazine*, Jan. 1860.

CHAPTER V.

FORMATION OF ICE.—ICE-FLOES, FIELDS OF ICE, AND ICEBERGS.—ICE IN THE BALTIC AND THE BLACK SEA.

In the Polar seas the low temperature results in the formation of ice. During the long winters of these cold regions, the tranquil water of the bays and gulfs freezes round the edge of the coasts, and the crystallized mass gaining incessantly on the sea finally extends to a considerable distance. This is "ground-ice." The surface of the sea disappears like that of the lakes under a solid layer; but the manner of forming the icy crust differs, for in the rivers and basins of fresh water, crystals of ice at first appear over almost the entire surface, but in the seas which have no great depth it is generally from the bed itself that the liquid mass congeals.

In fact, salt water has not, like fresh water, its greatest density at the temperature of 39·2 degrees Fahr., but it becomes heavier and heavier in proportion as it freezes. The coldest strata of water, being also the heaviest, descend vertically towards the bottom of the sea, and displace the lower strata, which are lighter and of a higher temperature. While the water which descends to the bottom in rivers has a normal heat of seven degrees above freezing-point, the sea-water which falls deeper may have been chilled to 32° Fahr., or even many degrees below it. When the mass is not agitated, it remains liquid, but, on the slightest disturbance, it suddenly turns to ice. Sometimes, at the commencement of winter, the mariners and fishermen of the Baltic and western coasts of Norway find themselves suddenly surrounded by floes of ice, which rise from the bed of the sea and which still contain fragments of fucus. It appears so rapidly that the boats often run great risk of being crushed between the solid masses which are piled around them, and the crews are in imminent danger. Around the rocky coasts of Greenland, Labrador, and Spitzbergen, these ice-floes often raise huge stones which they have torn from the bed of the sea.[*]

[*] Edlund, *Poggendorf's Annalen*, cxxi.

In the open sea ice is also formed. In winter, when the air is calm, the snow falls in large flakes on the tranquil waves, the sea is soon covered with a kind of scum, which gradually changes into a thin coating of ice. The wind may break this layer when barely formed, and the tiny scattered fragments may be surrounded with water from the melted snow, which does not mix with the salt-water of the sea, and glitters feebly with iridescent hues beneath the rays of an oblique sun; but this does not last long, and the cold soon re-forms the layer of ice.* Even in despite of wind and wave, innumerable needles of ice, which give to the surface of the water a pasty appearance, spread their network over the sea, and soon consolidate into a thick layer, which constantly increases as the cold of winter becomes more and more rigorous. By the natural chemistry of the sea, which is an immense laboratory, the mass of ice is in a great measure freed from the salt which is found in sea-water; for, according to the observations of Mr. Walker, it contains hardly more than five thousandths; that is to say, about a fifth of its normal quantity. The water nearest to the new ice mixes with the expelled salt, becomes heavier, and as the freezing-point is at the same time lowered it descends deeper in the water without becoming solid. This is the reason why in the open sea the water is rarely frozen for any considerable depth below the surface, as one might expect.†

In consequence of the frequent collision of these fragments of ice tossed by the waves, they generally assume the same circular form as the flakes of ice on rivers. They are roundlets of a very inconsiderable diameter, slightly raised at the edges; the English sailors term them "ice-cakes." But the cold becoming more intense, these disks finally adhere to one another, and before long millions of them, united in vast fields, form islands which stretch to the farthest horizon. Sometimes these "ice-fields" have a superficial extent of hundreds of thousands of square miles, and even constitute, by their dimensions, real continents. Those which border upon the eastern coast of Greenland have not been melted for four centuries, and effectually prevent the approach of navigators to the land; those connected with the Siberian coasts are still more considerable, because of the long extent of shore which serves as their base. In the Polar archipelago of America, ice bars the entrances of the channels almost every year, and raises before the navigator an impass-

---

* Gustave Lambert, *Expédition au Pole Nord. Bulletin de la Société de Géographie*, Dec. 1867.

† Neumann, *Ueber die Dichtigkeit beim Meerwasser*.

able wall. How many times have the explorers of Arctic seas tried in vain to find a passage through these barriers, and have remained imprisoned in the solid mass, after having ventured into some deceitful opening of the ice-field!

These interminable white surfaces are almost always bordered on the seaward side by blocks and disks rocking or whirling on the billows; these are the scattered islands which announce the neighbourhood of continents of ice. Those which are elevated on an average from 3 to 6 feet above the water, and the bases of which descend from 20 to 25 feet below the surface, have sometimes a tolerable uniformity of aspect, and when at times the snow covers all inequalities the ice-field seems to be transformed into an even plain like the Russian Steppes. But the ice is much more often rugged; fantastic hillocks, formed of all the wreck-fragments which the floes of ice have thrown up in dashing against each other, appear here and there several yards high. There are some which one might even confound with the enormous blocks that have fallen from the glaciers of Greenland or of Spitzbergen, and which really cannot be distinguished from them but by the slightly saline taste of the ice. These projecting masses are seen from afar above the sea, and remain erect long after the ice-field has melted. In the Siberian seas, where they give them the name of *toroses*, most of these hillocks, composed of the ice of the preceding winter, are easily melted by the first warmth of summer; but there are some which are preserved from year to year, and which remain indestructible during centuries, even under the rays of the sun. The Ostiac hunters, who frequently see these *toroses* run aground on the Siberian coast, designate them "Adam's ice," and imagining that they are contemporaneous with the origin of the world, assert that even fire itself is powerless against their crystalline masses.*

In spring-time and in summer, when the great heat commences in the polar zone, the force of the currents, whose action constantly makes itself felt beneath the ice plains, detaches from the remainder of the mass enormous fields of ice several hundred square miles in extent, and carries them far towards the open sea. The vessels of the explorers or whalers, which have been set fast in the bed of ice, are then carried out of their course with the broken field. Courageous sailors who have penetrated beyond Baffin's Bay have often thus been brought back by the current for hundreds and thousands of miles, and have only been able to regain the way they

* F. de Wrangel, *Voyage, Appendice*, p. 314.

have lost at the price of most painful efforts, or have even been obliged to abandon their enterprise completely. Such was the case in the sea around Spitzbergen in 1777; ten Dutch vessels were driven with the ice more than 1500 miles towards the southwest, and shattered on the way. It was a phenomenon of the same nature which prevented Captain Parry from reaching the North Pole. He had already approached nearer to this point than all preceding navigators, and had taken a sledge to cross the icefield; but each day, notwithstanding the great distance apparently traversed in the direction of the pole, he found himself further than the day before from the goal towards which he marched. The reason being that the continent of ice which bore him was being itself carried rapidly towards the south. White bears are thus sometimes carried by ice-floes, and landed on the coasts of Lapland.*

When once broken, the ice-field soon disappears; large fragments driven by the currents and the waves are dashed against each other with the enormous force which a weight of hundreds of thousands or millions of tons gives. Shattered by the terrible shock, these masses are divided into pieces of smaller dimensions; the cementing ice being destroyed by the fragments of the more anciently-formed icefield, the turrets and pinnacles which stand here and there begin to melt and fall, and a few days after the thaw has commenced nothing remains but a few ice-floes and uneven blocks gently rocking with the waves. To account for this rapid disappearance of the icefields (in which the infinite tiny organisms† of the sea also aid) the inhabitants of Greenland imagine that the entire mass is engulfed in the depths of the ocean. Even in the Baltic, where this phenomenon is comparatively much less remarkable, the Danish sailors, almost without exception, assert that in spring-time the ice-floes are swallowed up by the sea, although not one of them has witnessed the immersion.‡ But what is more easily corroborated is the strange noise that always accompanies the breaking up of the ice. With the crash of the meeting ice, more deafening, more terrible than that of cannon answering to each other, with the roar of the waves, and the groaning sound from the breaking disks and the air which escapes from them, is joined a kind of crackling, similar to drops of rain falling on plates of metal. This noise, which is heard also on mountain glaciers,

---

* Barto von Löwenigh, *Mittheilungen von Petermann*, Ergänzungsheft xvi.
† See below, the section entitled, *Earth and its Flora*.
‡ Forchhammer, *Philosophical Transactions*, part I., p. 233. 1865.

FORMATION OF ICEBERGS.

results, as Tyndall has shown, from the incessant breaking up of the crystals which compose the mass.*

Whatever picturesque beauty there may be in the ice of marine formation constituting this field, it is far inferior to that of the masses which are detached from the glaciers of Greenland, Spitzbergen, and other countries of the North Pole. Enormous fragments may be separated from the end of the glacier in two different ways, according to the temperature of the sea into which they protrude. In Spitzbergen and on the coasts of southern Green-

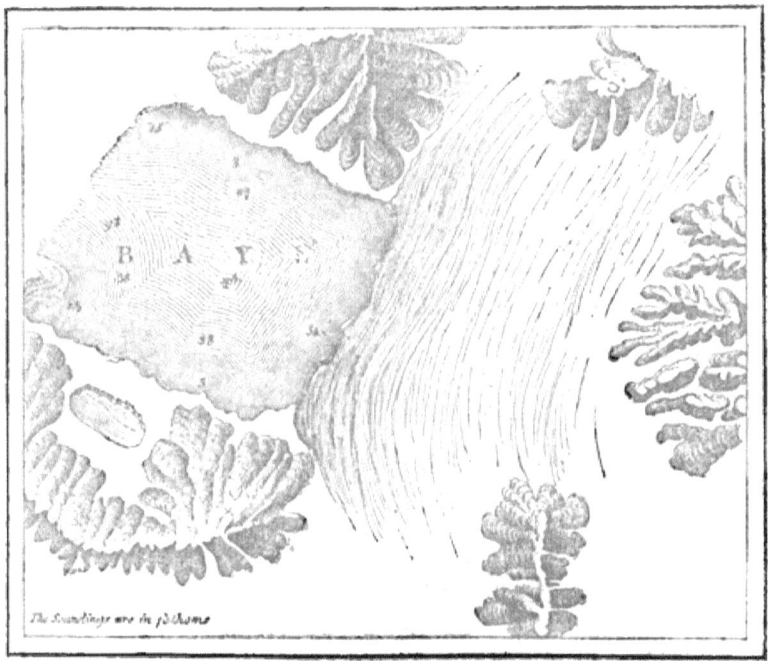

Fig. 12.—Glacier of La Madeleine, on the coast of Spitzbergen.

land, the congealed mass, which often projects far into the sea, is gradually undermined by the comparatively warm waves which beat against it, and the remaining fragments overhanging the water are detached with a terrible noise, and plunge into the ocean. M. Martins and other members of the French expedition to Spitzbergen have observed this at the base of all the glaciers of that archipelago.

* See in Vol. I., the chapter entitled, *Snows and Glaciers*.

But in very cold seas, like that of Smith's Strait, the water, being of a still lower temperature, cannot melt the glacier, which continues its course into the bay, its extreme end reaching far into the depths of the ocean, like an immense plane gliding over the rocks. Though lighter than the water, the enormous frozen mass is kept below because of its cohesion to the *mer-de-glace* which drives it along. But the moment comes when that connection breaks, and obeying at last the force which its buoyancy imparts to it, it shoots to the surface, and after repeated oscillations from the change in its centre of gravity it rises in huge towers or fantastic peaks.* We can imagine what a chaotic mass all these fragments, mixed with the marine ice and the remains of ice-fields, must produce in narrow bays, or in very contracted arms of the sea. It was across one of these prodigious "packs," in Smith's Strait, that the intrepid Hayes, with almost superhuman perseverance, passed.

These glistening icebergs are the splendour of Arctic seas. Often of colossal dimensions, they present at times forms of almost perfect regularity, whilst at others they assume the most varied and fantastic shapes. Lofty towers, columns in pairs, with groups of sculpture, and statues, like marble divinities, rise above the sea. In comparatively warm seas like those of Spitzbergen, which are affected by the Gulf-stream, the ice is constantly worn away; and those parts of the floating masses which rise above the surface of the sea generally assume the appearance of pillars, with more or less overhanging capitals, fringed with stalactites. The summit is white and occasionally covered with snow, while the fluting of the column where the more compact ice has been bathed by the waves has an emerald or sapphire hue. The foundations of the columns are pierced with caves, into which the water rushes with a hollow murmur; and at times they are riddled with small holes, from which each wave springs in diverging jets. Silvery fountains burst alternately from either side of the column according to the inclination given to it by the sea.† In very cold water, like that of the Arctic Archipelago, the opposite phenomena occur. Instead of being worn away and melted by the waves, the blocks fallen from the glaciers at first gradually increase in dimensions, on account of the low temperature of the water into which they are plunged, which solidifies around the foot of these enormous floating towers.‡

---

* Rink; Hayes, *The open Polar Sea.*
† Barto von Löwenigh, *Mittheilungen von Petermann*, Ergänzungsheft, xvi.
‡ Edlund, *Poggendorf's Annalen*, cxxi.

SIZE OF THE ICEBERGS.

The larger masses detached from the glaciers are known under the name of icebergs. Dr. Wallich was able to measure some of them on the coasts of Greenland, by ascertaining the depth below water of the bank on which several of these moving bergs had been stranded; and he found that, with the regularly formed blocks, the part above the level of the sea is never more than the fourteenth or sixteenth part of that beneath the level of the water. With respect to the masses whose exposed portions terminate in a cone or a pyramid, they descend to a less depth, in proportion as they present a more considerable bulk above water. But the total height of the iceberg always exceeds by seven or eight times the visible portion.

By these proportions, mariners can judge of the real size of the icy masses which they see stranded on the coasts of Newfoundland, or melting slowly as they float far out into the Atlantic. Enormous blocks have been seen from 300 to 400 feet high, so that these fragments of glaciers measured more than 3,000 feet from summit to base; that is to say, an elevation equal to the highest mountains of England or Ireland. One of these masses which was encountered by the ship *Acadia* off the bank of Newfoundland, amidst a labyrinth of other floating mountains, was about 480 feet high, surmounted by a kind of dome resembling St. Paul's Cathedral in a most singular manner. Twenty days later, when on her homeward voyage, the *Acadia* found the same iceberg 68 miles more to the south. A great number of these travelling masses have been seen, measuring several miles in length and breadth, whose bulk amounted to tens of thousands of cubic yards. As to fragments of ice-fields, some have been met with measuring not less than from 50 to 100 miles in each direction.

The slow movement of the block observed by the *Acadia*, which only advanced a little more than three miles per day, proves that icebergs offer considerable resistance to the current which carries them. The checks to which they are subject on the way, such as partial strandings, or when the surface and under currents urge them in opposite directions, retard their speed considerably, and often change them into seemingly stationary islets. Towards the end of 1855 an unexpected circumstance, still more remarkable than that of the berg seen by the *Acadia*, shows us exactly what had been the progress of an iceberg during the space of more than a year. An American whaler sailing in Davis's Strait perceived a dark mass in the middle of a group of floating peaks; this mass was the ship *Resolute* which the British Government had sent out in search of Franklin, and which the crew, having ventured into the ice-pack, had abandoned to continue their way in sledges. When the vessel was found again, it had been already detained in its floating prison for sixteen months, and during that space of time had only been carried about 870 miles, counting the necessary turnings through Barrow's Strait and Lancaster Sound. Thus the ship, abandoned in the Polar sea, had not exceeded the speed of 130 yards per hour in its progress towards the Atlantic, which is a hardly perceptible advance. In the history of the great Arctic expeditions three other vessels are mentioned, which were carried in the same manner towards the ocean, but without having been abandoned by their crews; these were the ships of Sir John Ross, of Lieutenant de Haven, and of McClintock. The last-named navigator was a prisoner for 242 days, and advanced about 1120 miles towards the south, that is to say, about 346 yards per hour.

The enormous masses of icebergs like gigantic ships are often stranded on shoals, even where the depth of the sea exceeds a hundred fathoms. Arrested in its southward drifting the immense block gradually dissolves or divides into fragments, which in their turn are stranded on some other bank, at a less depth. Day by day the waves melt and destroy great quantities of ice, which then let fall the gravel and stones with which it was charged, and in this manner continually raises the sea-bottom. Every year new beds of rock, pebbles, and earth from the mountains of Greenland and the archipelago of North America are thus deposited on the bank of Newfoundland, and in the neighbouring seas, laying the foundations of a new continent. Doubtless the Great Bank, which extends over a tract of above 55,000 square miles, and which has its foundation in a sea of about four to six miles deep, is composed entirely of this moraine-matter of glacial origin. Thus during a long series of ages the ice-floes have been labouring without relaxation to demolish the Arctic lands, and to construct new continents in the seas of the temperate zone.

From the time of the breaking up of the northern ice, that is to say, from the beginning of March to the month of July and even to the month of August, that part of the Atlantic to the east of the bank of Newfoundland assumes the appearance of the Arctic sea. The Polar current, descending from Baffin's Bay parallel to the coasts of Labrador, brings with it in long procession the fragments of the ice-fields and glaciers of Greenland. After having rounded the bank of Newfoundland, the current bends towards the southwest with

DANGERS FROM ICEBERGS. 43

its burden of ice, in consequence of the movement which carries the earth in an easterly direction and causes a deviation from its course in everything coming from the north.* Carried by this current, which drives them in the opposite direction to the Gulf-stream, continuing its course towards the south-west below the surface current

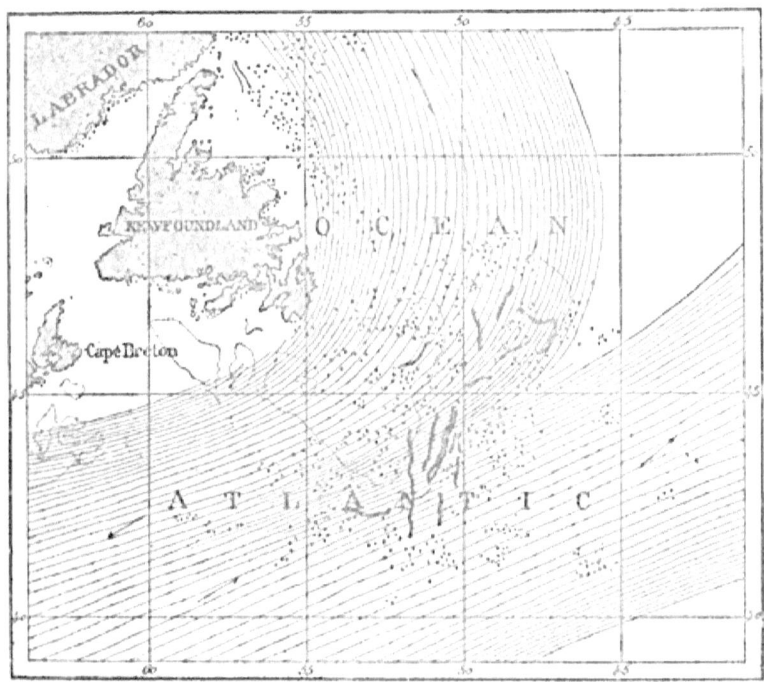

Fig. 13.—Course of the Icebergs between Europe and America.

of the latter, the icebergs, like ships cutting the waves with their prows, pass majestically through the water which dashes against them. Some fragments of mighty ice-fields, brought from Greenland by Polar currents and then drifted northward by the Gulf-stream, are seen here and there proceeding in an opposite direction from the rest. The accompanying map, borrowed from Redfield, indicates the position of all the icebergs and ice-fields recently observed in the western part of the North Atlantic Ocean.

It is principally in this region of the ocean that flotillas of ice are to be dreaded by navigators. The sailors of Newfoundland hardly

* See below, p. 62.

ever approach one nearer than about a mile, and then always keeping to windward of them, for otherwise they would be in danger of drifting upon the terrible mass, towards which in addition a somewhat strong current is always flowing to replace the upper stratum of water rendered colder by contact with the floating mountain. Enveloped in fog in consequence of the lowness of their temperature compared with that of the warm humid air from the south, the gigantic hull of the glacier discovers itself to seamen by strange whitish reflections and also by the intense cold of the surrounding atmosphere. But sometimes when this indication of peril has just been recognized it is too late to avoid the shock. Hundreds of ships overtaken by the ice have thus disappeared with their crews in the cold waters of the ocean. At other times, even in clear weather, one meets with a whole archipelago of ice-floes; and in order to avoid them it is necessary to steer with the greatest precaution for days together. It was thus that in 1821 the English brig *Anne*, surprised by the ice before Cape Race, not being able to enter a free sea, was obliged to remain 29 days surrounded by towers and threatening peaks. Happily these fragments of glaciers diminish very quickly in number and height as soon as they enter the zone of the Gulf-stream. Worn away at their base by the tepid waters of that current they capsize, break, and dissolve so completely, that towards the 40th degree of latitude it is rarely that any fragments even remain. However in June, 1842, the ship *Formosa* encountered in 37° 30' of north latitude a floating iceberg 30 yards high and 50 yards long moving towards the south.*

In the Antarctic hemisphere, exactly similar phenomena occur. Thus, as is proved by numerous observations, more than 860 of which have been regularly catalogued by Fitzroy and other geographers, the ice-fields and fragments of glaciers of the southern continent float likewise in the direction of the equator. But it seems that the icebergs of the southern hemisphere generally present less variety of form than those of the opposite one. They are not peaks and domes with fantastic outlines, but rather resemble walls rising like rocky precipices to an elevation of about 160 to 200 feet; these floating masses are probably, however, on an average of still more considerable dimensions than the masses which fall from Arctic glaciers. The massive form of these floating mountains of the southern seas must doubtless be attributed to the severe cold which prevails in the south polar zone, which drives the snow and

* Redfield, *Memoir on the dangers of ice in the North Atlantic Ocean.*

glaciers of the Antarctic lands further into the open sea. Even at the 50th degree of south latitude, ships meet with ice-fields of a size equal to those which on the other side of the earth are only found within the polar circle. In the northern hemisphere the ice-rivers of Greenland and Spitzbergen are not fed by a sufficient quantity of snow to carry them completely out of the bays into which they flow, and into the open sea. Retained in their course by steep lateral cliffs, promontories, and rocky islets, they assume in consequence of all these obstacles a much more irregular form than they would have if they penetrated into the free ocean, like the glaciers of the South

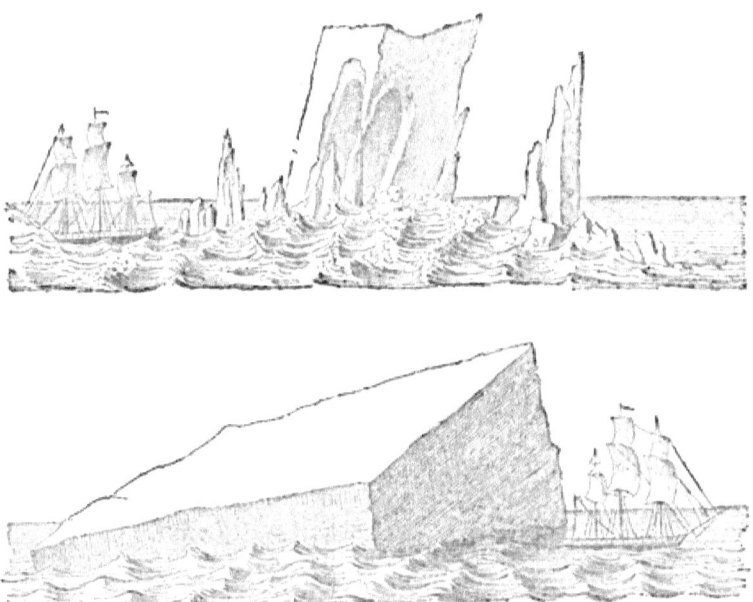

Figs. 14 and 15.—Icebergs of the Antarctic Ocean; after Wilkes.

Pole. The latter are drifted far out of the gulfs, beyond the capes even, and they are only occasionally attached to the submarine base of the continent. In front of this ice-sheet float innumerable islands, through which ships can with difficulty find their way. Thus during the exploring voyage of Wilkes, the *Peacock* had to steer for a long time in a labyrinth of blocks which threatened to crush her.

The breaking up of the Antarctic ice occurs in spring and summer, like that of the North Pole, but six months later, in consequence of the opposition of seasons in the two hemispheres, caused by the obliquity of the earth's axis. The scattered pieces of ice met with during winter are only fragments detached from the ice-fields.

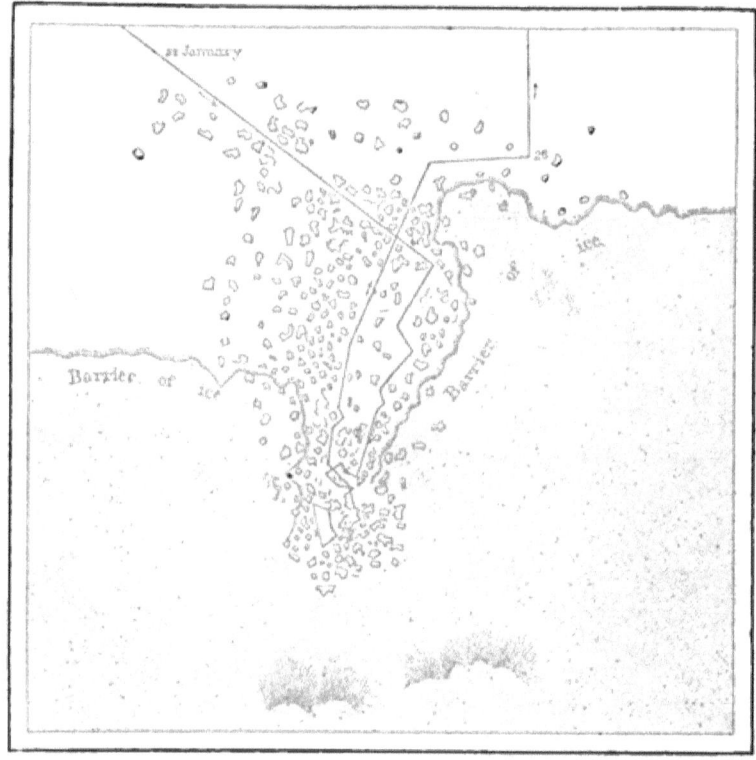

Fig. 16. Rou e of the *Peacock*; Commander Wilkes, U. S. Navy, in the Antarctic Ice-pack.

Vessels traversing the Antarctic Ocean meet with thirty or forty times more ice in December, the height of summer, than in July, which is the coldest time. The multitude of floating masses varies much in these seas. To the south of Australia and New Zealand icebergs and ice-fields are comparatively rare. To the south of Cape Horn they are met with more frequently, but are never seen between this southernmost point of America and the Falkland Islands; for owing to the great Polar current they all drift towards the northeast. It is to the south of the African continent that the ice is

The Ocean &c   ANTA

P A C I F I

Knox's Land   Antarctic   Circle

Engraved by Erhard

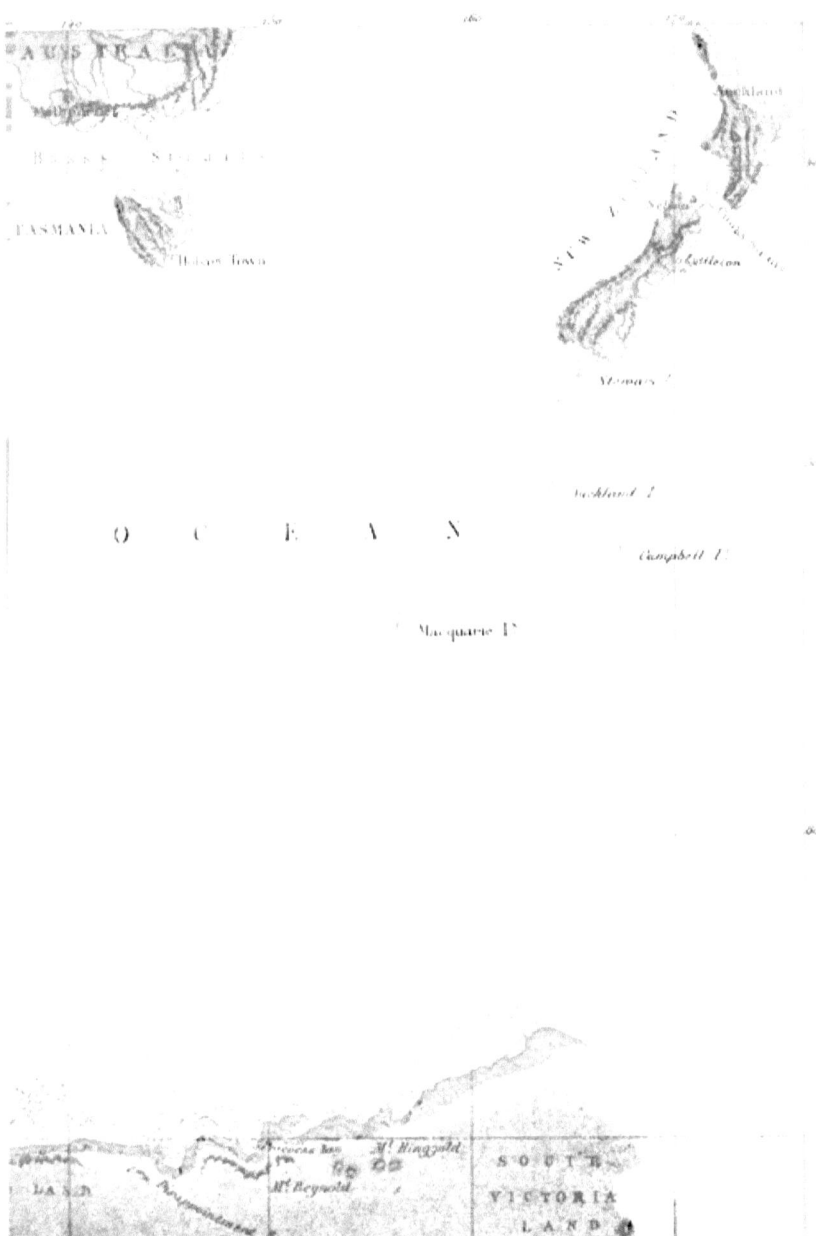

Drawn by A. Vuillemin. — after Ch. Wilkes

carried in the greatest quantity, and approaches most nearly to the equator. Some has even been perceived from Cape Town in 34

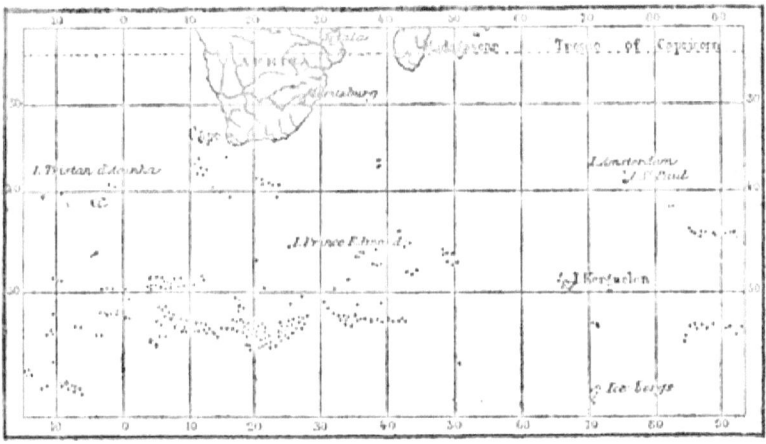

Fig. 17.—Course of Icebergs in the Southern Hemisphere.

degrees of south latitude. Thus the Antarctic icebergs are carried about 250 miles nearer the torrid zone than are the Arctic masses.

In the inland seas exposed to severe cold, the congelation of the water is produced in the same way as in the ocean; the phenomena only differ in proportion. Thus the ice of the Baltic is far from presenting such a grand spectacle as the ice-fields of the Polar seas, but its mode of formation is known in a much more complete manner; for during a long series of years, conscientious observers have studied its various changes, from the formation of the first ice to the general breaking up. These researches have proved that after having been formed, the icy bed of the Baltic is subject to the same phenomena as that of lakes, not only in the northern parts of the sea where the water is almost fresh, but near the entrance also where the mass of fluid is still strongly saline. The crevasses in the ice do not differ essentially in their formation from those of Lake Baikal [*] or the Lake of Constance. They also open with a thundering noise letting a great quantity of water escape, which freezes in its turn and thus increases the thickness of the solid bed. Around the island of Oesel the fissures vary from six inches to more than six feet, and are continued for a

[*] See in Vol. I. the section entitled, *Lakes*.

distance of several miles. But the surf produced by the currents and the dashing of the waves, where the sea is not frozen, gives the most varied directions to the crevasses; in some places they are parallel, while in others they intersect one another irregularly or radiate towards all points of the horizon.

Ice very rarely covers the surface of the sea while the water is much agitated. Tempests or rapid currents retard, or even completely prevent, the formation of the ice-sheet. Thus, while on the east, where the sea is calm, the island of Oesel is, on an average, united to the mainland during 130 days of the year by a layer of ice sometimes attaining a thickness of more than three feet, and serving as a high road for sledges; the western cliffs, against which the surges strike, are, on the contrary, only bordered by a narrow fringe of ice. On the promontory of Muhha Ninna the waves always break with fury, and this extreme agitation of the water lasts during the whole winter, preventing the appearance of the least particle of ice; indeed the peasants of the island say that they have never seen any near this point.*

Every year, a considerable part of the Baltic is covered with ice. Almost all the Gulf of Bothnia and the entire coast-line of the Gulf of Finland is changed into a white and immovable surface, the islands and islets are encircled by a zone of ice-floes, more or less wide, whilst the straits of a slight depth are similarly obstructed. Every winter Finland is reunited to Sweden by a bridge of ice, pierced here and there by the innumerable rocks of the Oeland archipelago. This solid crust then becomes for many months the highway between Sweden and Russia. The Baltic, like the Polar ice-fields, has its piled-up masses of ice resembling turrets, pyramids, and obelisks built upon the sea; from these fields, also, broad masses are detached from their edges to float towards the south with the current, then breaking with a loud crash, are similarly reduced into scattered pieces; and in a few days after the commencement of the thaw only thin fragments remain, tossed here and there by the waves.

During the last few centuries the Baltic Sea has never been entirely covered with a field of ice. But the chronicles inform us that in 1323 the southern part of the basin was completely frozen over, and during six weeks travellers from Copenhagen repaired on horseback to Lubeck and Dantzic; and temporary hamlets were even erected on the ice at the intersection of the roads. During the winters of 1333, of 1349, 1399, and 1402, the same phenomena of

* Von Sass, *Bulletin de l'Académie de Saint Petersbourg*, t. ix. p. 166, &c.

general congelation occurred in the southern Baltic, and the icy bed served as a road for commerce between Pomerania, Mecklenburg, Denmark, and the islands. In 1408 the ice-field completely closed the entrance of the Baltic between Norway and Jutland, and extended through the Cattegat, the straits of the sea of Scania, into the Baltic, as far as the large island of Gothland. It is said even that the wolves of Norway, driven from their native forests by hunger, crossed the Skagerrack to invade the villages of Jutland. Since this epoch, several parts of the southern Baltic have been frozen over again; but the solid surface has never presented the same extent, nor the same consistency. This fact would seem to prove that the mean temperature has become milder in Northern Europe since the 14th century, while according to Adhémar's hypothesis exactly the contrary is the case.*

It is a remarkable fact, that save in a few exceptional years the Black Sea, which is exposed to all the piercing winds which descend from the Polar regions, has never been invaded by ice like the Baltic. During the earlier historic ages the Sea of Marmora and the surface of the Euxine have been frequently covered with ice; which proves that, at least during this period of frost, the temperature of Constantinople was no higher than that of Copenhagen. In the year 401 of the present era the Black Sea was almost entirely frozen over, and when the ice broke up, enormous icebergs were seen floating in the Sea of Marmora for thirty days. In 762 the solid layer which covered the Euxine extended from one bank to the other, from the terminal cliffs of the Caucasus to the mouths of the Dniester, Dnieper, and the Danube. Moreover, contemporary writers assert that the quantity of snow which fell on the ice rose to the height of twenty cubits (from 30 to 40 feet?), and completely hid the contour of the shores, so that one knew not where the land began or the sea ended. In the month of February, the broken masses of the ice, carried by the current to the entrance of the Ægean Sea, reunited in one immense sheet between Sestos and Abydos across the Hellespont.†

* See in Vol. I., the chapter entitled, *Harmonies and Contrasts*.
† P. de Tchihatcheff, *Le Bosphore et Constantinople*.

CHAPTER VI.

WAVES OF THE SEA.—REGULAR AND IRREGULAR UNDULATIONS.—HEIGHT OF THE WAVES.—THEIR SIZE AND SPEED.—GROUND-SWELL.—COAST-WAVES.

The sea rarely presents a glassy surface. When the atmosphere is calm, which however is commonly the case before a tempest, the water is sometimes so very smooth that every object is reflected by it with a perfectly sharp outline; the only changes which seem to affect the vast motionless sheet of water are those produced by the mirage, which makes the distant horizon glitter like a long band of silver or steel; the fishermen then say that "the sea is reflecting itself." But this tranquillity of the water is a very uncommon phenomenon, except in the Mediterranean and other seas, where there is only a slight tide. Usually the wind, either in breezes or tempests, now aiding and now retarding the ebb and flow, raises the sea into waves, more or less high, which sometimes roll onward regularly, or are dashed against and cross one another. Even during calms, the waves, still obeying the impulse of recent winds, continue to roll across the ocean in long undulations. One of the grandest spectacles at sea is offered by these regular movements of the waves in perfectly calm weather, when not a breath stirs the sails; high, blue, and foamless the liquid masses succeed one another at intervals of 200 to 300 yards, pass under the ship in silence, and pursued by other waves are lost in the far distance. One contemplates with a feeling of admiration, not unmixed with terror, the calm and majestic wave advancing like a moving rampart, as if about to swallow up all before it, and yet hardly leaving a sign to mark its passage. These waves appear with surprising regularity, during the autumnal calms, under the tropic of Cancer, and almost at every season in the narrower part of the Caribbean Sea towards the Gulf of Darien; there the waves are seen silently to advance, and slightly raise the ship, passing onward with scarcely a murmur, as regularly as the furrows of a field, and stretching as far as the eye can see.

IRREGULAR UNDULATIONS.

Such perfectly regular waves as these can only be formed in seas exposed to the influence of equable winds, such as the trade-winds. Wherever the winds are uncertain and shifty, blowing in gusts, it is evident that the waves driven by them cannot assume a regular form or follow in a uniform direction. For aërial currents constantly vary in their speed; being composed of strata of unequal force, which moving at a rate different from that of the surface of the sea, alternately increase and diminish in force. Under the influence of these variable atmospheric impulses, the waves must necessarily vary in height and speed, and their crests cannot be developed in a uniform line. The wind also frequently changes its direction; as if urged by some new impulse,* it commences blowing from another point of the compass, and drives the waves in a different direction from that which it had itself given them. Nevertheless the first movement is continued by the succeeding waves even while the second is still making itself felt, and from this double impulse an intersection of waves, differing from one another in direction, height, and speed, results. Let the wind shift to another point of the compass, and a third undulation crosses the preceding two. Finally, should the aërial current make the complete circuit of the compass, the ripples of the water pursue one another in all directions, urged from all points of the immense circle. Not a breath is lost on the sensitive surface of the sea, and the variety of its undulations testifies to the diversity of the aërial movements which cause them.

From a lofty headland or from the mast of a ship, whence a vast expanse of water can be viewed, the beautiful sight may be often enjoyed of two or three systems of waves intersecting each other at various angles. Now they double the natural height of the undulations, by piling one wave upon another, and then again they equalize the surface of the water by throwing billows into the furrows. Sometimes the sea is so agitated that it is impossible to discern the direction of all the waves which have aided in producing the violent commotion. As to the voyagers, whom the wave-tossed ship incessantly shakes by its rolling and pitching, it is still more difficult for them to recognise in the intersection of the waves the various impulses communicated to the sea by the atmosphere. The accompanying figure is reproduced from that of an English traveller, of the curves drawn during a single minute by a pencil suspended vertically in the cabin of a ship. At the time when the pencil

---

* See below, the chapter entitled, *The Air and the Wind*.

traced these lines, the wind was low and the motion of the water very moderate.

Fig. 18.—Rollings of a ship upon the waves.

The height of the waves is not the same in all seas; it is greater when the basin is deeper in proportion to the exposure of its surface to the wind, and also in proportion as the water, being less salt and so lighter, yields more readily to the atmospheric currents. Thus assuming equality of surface, the water of Lake Superior would be raised in higher waves than that of a gulf of the sea barred on the open side by islands and sand-banks. When of equal saltness, the narrowest basins ought to present the shortest and least elevated waves. The waves of the Caspian Sea are not to be compared with those of the Mediterranean, which again are greatly exceeded in height by those of the North Atlantic; and these latter in their turn are surpassed by those of the Antarctic Sea, which spreads over an entire hemisphere.

According to Admiral Smyth, who was well acquainted with the Mediterranean, the tempest waves rise from 13 to 18 feet in vertical height above the trough of the sea. He has even seen quite exceptional waves rise to the height of above 30 feet, but the average waves raised by high winds were only from about 10 to 13 feet.*
In one passage from Liverpool to Boston, which the celebrated

---

* Cialdi, *Sul moto ondoso del mare*, p. 142.

navigator Scoresby made in 1847, he measured waves from 26 to 29½ feet, and the average of all his observations gave a height of about 19 feet for the largest waves. On his return in 1848, he found the average to be 30 feet, and some among the waves he measured rose to about 43 feet above the trough of the sea. Other navigators have given similar estimates for the highest crests of waves in the North Atlantic; but the mean elevation is much less. One can form a good notion by the following diagram, drawn by the engineer Middlemiss to represent the annual variations of the wave at Lybster on the coast of Scotland.

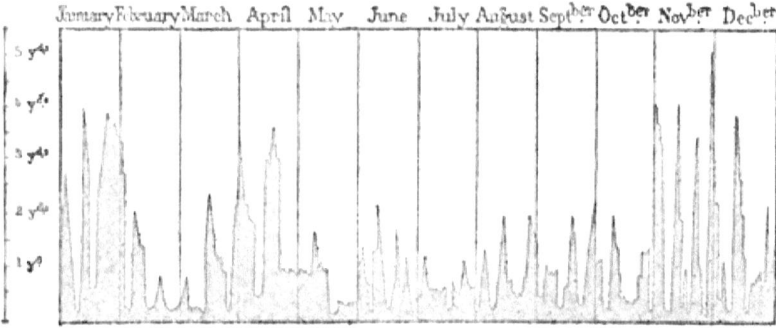

Fig. 10.—Average heights of Waves observed at Lybster (Scotland) in 1852.

In the South Atlantic the height of the waves is certainly greater than in the northern parts. Many seamen have seen the water rise to between 50 and 60 feet off the Cape of Good Hope, where the basins of the Atlantic and the Indian Oceans meet. Dumont d'Urville even asserts that he has seen waves above 108 feet high, to the depths of which the ship descended as into a valley, and M. Fleuriot de Langle attests the truth of this assertion. These are indeed the mountains of which poets speak, and which, in fact, seem such to those who find themselves at their mercy. It is probable, too, that the highest waves of the sea have not yet been measured. One remarkable thing is, that it is not usually during the most violent tempests that the hugest waves are formed. On the contrary, the force of the atmosphere which then precipitates itself obliquely on the waves, so to speak, depresses and crushes them.*

The waves are developed in all their majesty when the wind is at the

* Cialdi, *Sul moto ondoso del mare*, p. 139.

same time very high and very regular, and blows for a long time from the same point of the compass.

As to the width of the waves, that is to say, their total breadth from base to base, observers have not obtained the same results; but there are few among them who have found the vertical height of the crest of the wave to be less than a twentieth or more than a tenth of the width. On an average the height of an undulation of the water is only equal to the fifteenth part of its base; thus a wave of 4 feet in height measures 40 feet from valley to valley, and a wave 33 feet high is 495 feet in width. This is a much smaller

Fig. 20.—Average Amplitude of Waves.

size than would be imagined by the sailor lost in the midst of the billows, which he sees rising around him in every direction. Moreover, the inclination of the waves varies with the force of the wind, and the movements of the secondary undulations which intersect the principal ones.

The speed of the waves is only an apparent speed, like that of the folds of a cloth, raised by a current of air. Thus, although the water pressed by the wind rises and sinks by turns, it nevertheless hardly changes its place, and objects floating on its surface move but slowly and in an undulatory manner. The real movement of the sea is that of a drifting current which gradually forms under the prolonged action of the wind; but this general movement of the liquid mass is after all inconsiderable. The only part which advances with the storm is the foaming crest which, curling over the summit of the waves, dashes down the slope in front. By their incessant movements, the surface of the waves gradually increase in temperature, as has been observed after a succession of violent storms.*

The apparent displacement of the billows (which is rather difficult to measure with exactitude in the open sea) varies in a regular manner, according to the magnitude of its waves and the depth of its waters. Thus according to the calculations of the astronomer Airey, every wave of 100 feet in width, traversing a sea of 164 fathoms mean depth, has a velocity of nearly 2100 feet per second,

* Joule, Cialdi, *Sul moto ondoso del mare*, p. 218.

or about 15½ miles per hour; a wave of 674 feet, moving over the surface of a sea 1,640 fathoms deep, travels more than 69 feet per second, or nearly 50 miles per hour; this last figure may be considered as an average speed for storm-waves in great seas. Since, therefore, we can by calculation infer the velocity of waves from their width and the known depth of the ocean-bed, it is easy to determine by an inverse operation what is the depth of the ocean itself, provided that we know the rate of motion of the waves. It is by this method that the mean depths of the South Atlantic and of the Pacific Ocean between Japan and California have been calculated.*

Natural philosophers have frequently discussed the question of the movement of the waves in a vertical direction. To what depth in the abysses of the sea does the action of the superficial wave penetrate, and at how many fathoms can it disturb the sand and debris at the bottom? It was formerly admitted, as a well-ascertained fact, but without proof, that the agitation of the sea ceases to be felt at 4 to 6 fathoms below the surface. Direct observations made by seamen in many latitudes have shown that this opinion is erroneous. Sailors have frequently seen the waves break at 10, 16, and even 27 fathoms of depth over hidden rocks, which proves that the rocks were obstacles which abruptly barred the advance of the lowest part of the wave. Still more frequently, during violent tempests, the water has been seen charged with clay and mud, which had been raised from the bottom at 50, 80, and even 100 fathoms below the surface.† The direct experiments of Weber on the movements of waves have likewise proved that each wave extends its influence in a vertical direction to 350 times its height. Thus every wave of about a foot in height makes itself felt on the bed of the North Sea at a depth of 50 fathoms; whilst every oceanic billow of 33 feet is felt at about 1¾ miles. It is true that at these enormous depths the action of the wave is, so to speak, imaginary, for below the surface it decreases in geometrical proportion; but at about 25 to 50 fathoms only, the submarine waves have still great force, and we can easily understand that when thousands of them are abruptly arrested by submarine rocks, and on the rapid slopes of sand-banks, violent eddies must be produced which afterwards returning to the surface of the water cause "heavy swells." From these causes arise those turbulent seas, which ships encounter at times in calm weather, especially in the

* See above, p. 17.
† Cialdi, *Sul moto ondoso del mare*, p. 174.

neighbourhood of submarine banks; also those "ground swells" which suddenly raise the surface of the sea and endanger boats; and those formidable tide-races which springing from the depths of the ocean advance abruptly upon its sloping beaches, destroying all they encounter on their way.

It is along the shores of continents and around rocky islands that ordinary waves and heavy surf appear in all their grandeur and assume dimensions truly formidable. In accordance with the more or less gradual inclination of the bottom to the shore, a wave coming from the open sea rolls over a bed more and more shallow, and must perforce slacken its speed; but at the same time it increases by its own depth the stratum of water which it overflows, and consequently the wave which follows it is subjected to less retardation of the impulsive force. The second wave constantly gains on the first, and finally reaches it, swelling its crest, and slackening its own speed in its turn, gives a third wave time to distance it also. Finally, near the strand, the liquid mass, swelled by the pursuing waves, and unable to spread out further at its base along the shore which is too near; it gains in height what it wants in breadth, and rising like a wall, it bends over with a wide curve in front, and breaks with a thundering sound, throwing water mixed with sand and foam far along the shore. This surge, which is dreadful indeed during tempests, rises much higher than the waves; to the ancients the whitening billows of the open sea, whose crests were seen to shine like the fleeces of sheep, were the flocks of Proteus; while the waves of the shore, still called in our days *cavalli* and *cavalloni* by the people of the south of Europe, were the foaming horses of Neptune.

The height to which the crests of some of these waves attain when the configuration of the coast favours the movement, seems sometimes to partake of the marvellous. The mass of water which rises vertically can then only be compared to an ascending cataract. Spallanzani relates that sometimes, in violent tempests, the waves reach half-way up, or even to the top, of Stromboluzzo, a peak of lava which rises near Stromboli 318 feet above the mean level of the sea. The Bell Rock lighthouse, which rises boldly to 112 feet in height on a rock off the Scottish coast, is often enveloped in waves and foam even long after the tempest has ceased to disturb the sea.[*] Smeaton, too, has seen waves covering the Eddystone lighthouse, and leaping in a spout of water 82 feet above the lantern; the mass which is thus raised around the edifice cannot be less than from 2616 to 3924 cubic yards,

[*] Mrs. Somerville, *Physical Geography*.

FORCE OF THE WAVES.

and would weigh as much as a large three-decker. After these great storms, salt pools are scattered here and there on the top of the cliffs.

The pressure exerted by these masses of water, hurled with such impetus, is no less surprising. Thomas Stephenson ascertained that the force of the sea dashed against the Bell Rock lighthouse amounted to about 17 tons for every square yard. In the island of Skerryvore the heaviest calculated pressure is about 3 tons and a half for every yard, that is to say, more than $6\frac{1}{2}$ lbs av. for every 0·16 of a square inch. With such a force the displacement of blocks which seem enormous to us, is only child's play to the tempest waves. Before all sea-ports, and roadsteads where great works, such as sea-walls and breakwaters, have been constructed, seamen have been able to observe the prodigious power of the angry water. On all the exposed works at Holyhead, Kingston, Portland, Cherbourg, Port Vendres, Leghorn, the waves have been seen to seize blocks weighing several tons, and hurl them like playthings over the dikes. At Cherbourg the heaviest cannon on the rampart have been displaced; at Barra Head in the Hebrides, Thomas Stephenson states that a block of stone of 43 tons was driven more than $1\frac{3}{4}$ yards by the breakers. At Plymouth, a vessel weighing 200 tons was thrown without being broken to the very top of the dike, where it remained erect as on a shelf beyond the fury of the waves. At Dunkirk, M. Villarceau has ascertained by the most delicate measurements, that during a heavy sea the ground trembles at nearly one mile from the shore.

In the Gulf of Gascony, so frequently visited by tempests, the waves, coming from the west and north-west, are drawn into a sort of funnel, and hurl themselves against the shores with a force at least equal to that of the waves in the Channel, and the English seas. The works constructed by engineers to protect the roads and ports against this terrible pressure have been frequently swept away, or much damaged by the waves. Man must incessantly continue the strife he is engaged in with the sea, under pain of seeing the work of a century destroyed in a day. During the winter of 1867 and 1868, M. Palaà says that blocks of masonry, 36 tons in weight, placed at the extremity of the dike at Biarritz, were thrown horizontally from 11 to 13 yards; one block was even raised nearly 7 feet, carried over the breakwater, then thrown down, and rolled to a great distance during the storm. At St. Jean de Luz the surge is perhaps still more terrible, and some of the masses of stone now employed in constructing the dike of Socoa, at the entrance to the roadstead, are not less than from 80 to 90 cubic yards. And yet even this strong wall would not be strong enough, if it was not additionally defended by stones scattered loosely here and there, forming a range of protecting rocks in front of the dike upon which the sea expends its fury.

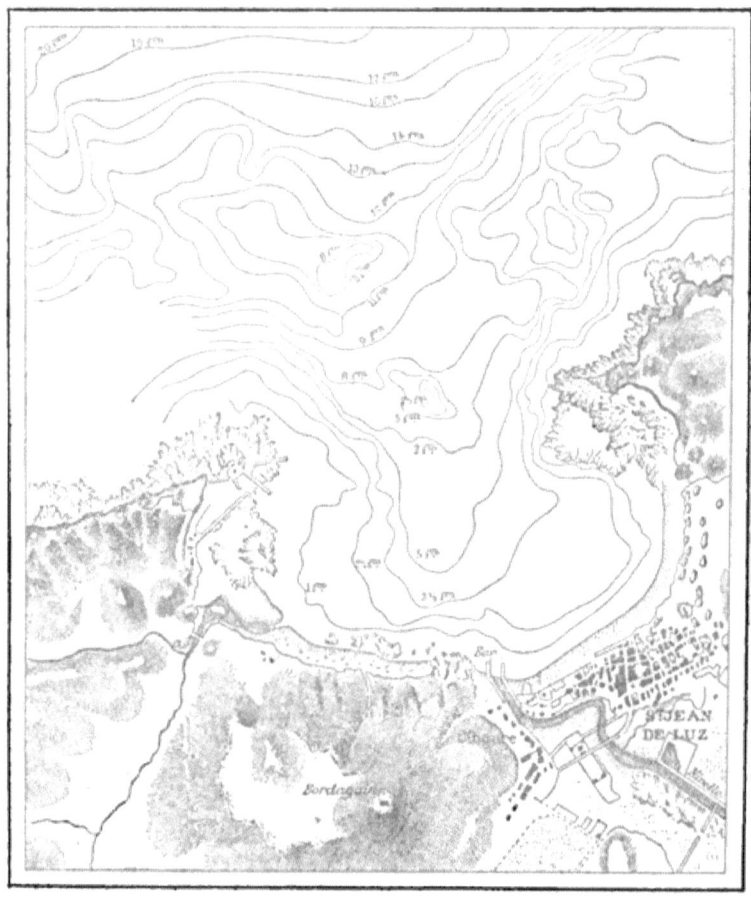

Fig. 21.—Bay of Saint Jean-de-Luz.

The only places where the waves display a still greater power than in the Gulf of Gascony, are those that are sometimes ravaged by the tornados. In the Isle of Réunion there is to be found in the middle of a savannah, a massive block of madreporic stone, which is

no less than 510 cubic yards in size. It is a piece that the waves have detached from a reef and driven before them across the land.*
How can we wonder that waves strong enough to hurl such projectiles can alter the shores in such varied ways, demolishing the cliffs here, and forming islands there, or constructing sand-banks at the entrances to gulfs.†

* Zurcher and Margollé.
† See below, the chapters entitled, *The Shores and the Islands; The Work of Man.*

BOOK II.—CURRENTS.

CHAPTER VII.

GREAT MOVEMENTS OF THE SEA.—GENERAL CAUSES OF CURRENTS.—
THE FIVE OCEANIC RIVERS.

CURRENTS, that is to say, the real movements of the sea, much less visible to the eye than the apparent displacements which constitute the waves, are notwithstanding of much greater importance in the economy of our planet. By their action enormous volumes of water, thousands of miles wide, and hundreds of fathoms deep, move across the oceanic basins; the water of the polar seas is carried to equatorial regions, while these on their side send their waves in the direction of the poles. The liquid mass circulates incessantly, as if in a vast whirlpool, in every ocean of the globe, and we can follow in thought its gigantic circuit from the fields of ice to the warm atmosphere of the tropics. Currents are indeed only the ocean itself in motion, and by their action the waters of the sea are successively distributed over all parts of the globe. They are the windings of the great "salt river" of Homer, which rolls around the earth in one immense circuit. Every drop that has not already been raised in vapour to commence its long journey through clouds, mists, glaciers, and rivers, continually changes its place in the abysses of the sea; it descends to the bottom, or mounts to the surface; it moves from the equator to the pole, or from the pole to the equator; and thus traverses all parts of the ocean. It is to this continual displacement of its innumerable particles that the sea owes its uniformity in such a surprising manner, under all latitudes, as regards the appearance, composition, and saltness of its waters.

Every difference of level which is produced on the liquid surface in consequence of prolonged winds, heavy rains, or very active evaporation, causes, as a necessary result, the formation of a current; for

GENERAL CAUSES OF CURRENTS.

water, whether salt or fresh, ever seeks its level and incessantly flows from the more elevated places towards the depressions. Every atmospheric variation has, for result, a displacement in one direction or another of the superficial water. But the great currents which flow with a regular movement around the basins of the ocean, between the polar and the equatorial zones, are determined by general causes acting at the same time on the entire planet. These causes are the sun's heat and the rotation of the earth on its axis.

The equatorial basin, incessantly heated by the solar rays, loses a great quantity of water, which is transformed into vapour, and rises into the higher strata of the atmosphere to be condensed into clouds. Admitting that the annual evaporation is about 14 feet,* which is probably below the reality, the quantity of fluid raised from the Atlantic in the tropical zone would be nearly 120 trillions of cubic yards, and would consequently represent the same value as a cubic mass of water nearly 30 miles in extent. It is true that a considerable part of this vapour, the half perhaps, falls as rain into the sea from which it was taken, yet a great proportion of the clouds are carried by the trade-winds † and other aërial currents, into seas situated beyond the tropics, and over the neighbouring continents. Near the equator therefore much more water is drawn from the ocean by evaporation, than is returned to it by the clouds, and in consequence an immense void is formed which can only be filled by the waters from the polar basins, where the contributions of snow, rain, and ice exceed the loss in vapour. This superabundant mass of fluid continually flows towards the basin of the torrid zone, and forms the two great currents, which meet one another from the opposite poles in the Atlantic and the Pacific, incessantly describing a regular orbit like the celestial bodies. But the excess of evaporation which occurs in tropical waters is not the only reason of this great movement of the polar seas towards the torrid zone. The trade-winds, attracted by the force of equatorial heat, blow incessantly in the same direction, and always driving the waves before them thus accelerate the march of the oceanic current.

If the mass of water which continually flows from the poles to the equator were exactly equal in quantity to that which is evaporated by the sun's heat, the arctic currents would be arrested under the tropics, and no return movement would be produced towards the

* Maury's *Geography of the Sea.*
† See below, the chapter entitled, *The Air and the Winds.*

polar oceans. But the waters which flow from the north and south are always in excess, in consequence of the continual impulse of the trade-wind; and when they arrive in tropical latitudes they are influenced by a new current, the true cause of which is the rotation of the earth on its axis. In fact, owing to the incoherence of its particles, the ocean does not obey in an absolute manner the rotatory motion of the earth, which carries it from west to east. In descending from the poles to the equator, and thus crossing latitudes whose speed of rotation is greater than their own, they are constantly drawn obliquely towards the west, and this continual retardation of their motion behind that of the rotation of the globe becomes, in relation to the surface of the sea, an apparent motion from east to west. Upon their meeting in the tropics, the polar currents, being both affected by a side movement, strike each other obliquely, then re-unite in the same oceanic river, and flow directly towards the west in the opposite direction to that of the solid earth. It is thus that the equatorial current is produced, which with the two polar currents determines all the movements of the waters in each oceanic basin. The other rivers of the sea are simply branches from them, caused by the form of the continents.

The equatorial current, which is a continuation of the polar currents and forms with them a vast semi-circle, cannot be freely developed around the circumference of the globe. Arrested in the Atlantic by the American continent, in the Pacific by Asia and the archipelago which unites that continent with New Holland, it breaks against the shores and divides into two halves which flow back in the direction of the poles, the one descending towards the south, the other ascending to the north. The immense river thus returns to its source, but at the same time the motion of terrestrial rotation, which at its outset caused it incessantly to deviate towards the west, now urges it obliquely in the opposite direction. Under the equator, the angular speed of the terrestrial surface around the axis of the planet being much more considerable than under any other latitude, the waters coming from the tropics into temperate seas are animated by a more rapid movement towards the east than those amidst which they flow. They deviate in consequence in an easterly direction, and when the returning current reaches the polar sea it seems to come from the west. Thus the grand circuit of the waters is completed in each hemisphere. The Atlantic and the Pacific have each their double circulatory system, formed of two immense eddies united in the torrid zone by a common equatorial current. As regards the Indian Ocean, being bounded on the north by the continent of Asia, it has but one simple current, which turns incessantly in its vast basin between Australia and Africa. As a whole, these ocean rivers recall by their distribution the divisions of the land. The two great whirlpools of the Atlantic correspond to the two continents of Europe and Africa; the huge eddies of the Pacific have a binary division analogous to the two continents of America; and the current of the Indian Ocean reminds one of the enormous mass of Asia, which alone fills half the northern hemisphere.

THE OCEAN.

CHAPTER VIII.

THE GULF-STREAM.—INFLUENCE OF THIS CURRENT ON CLIMATE.—ITS IMPORTANCE TO COMMERCE.

Of all the oceanic rivers, the best known to us is that part of the North Atlantic current, which the English and the Americans have named the Gulf-stream, because it makes a long circuit in the Gulf of Mexico before reaching the ocean. In the year 1513 the Spaniards Ponce de Leon and Antonio de Alaminos knew of the existence of this current; and six years later Alaminos, setting forth from the Straits of Florida, allowed himself to be carried by the water into the open sea, and thus discovered the great circular route which ships have now to follow in order to return speedily to Europe. Since the time of Varenius who attempted to describe the Gulf-stream, of Vossius who traced its immense circuit on a map, Franklin and Blagden who were the first to explore it scientifically, this current has been studied by numerous geographers. Without doubt, there is no marine current which merits to be better known in all its details; none has been of more importance in the commerce of nations or exercises a greater influence upon the climate of the North-West of Europe. It is to the Gulf-stream that the British Isles, France, and the neighbouring countries owe in great part their mild temperature, their agricultural wealth, and in consequence a very considerable part of their material and moral power.* Its history is almost identical with that of the entire North Atlantic Ocean, so important is its hydrological and climatic influence.†

The celebrated Maury devotes the most important part of his classical work on the *Geography of the Sea* to the Gulf-stream. It "is a river in the ocean; in the severest droughts it never fails, in the mightiest floods it never overflows. Its banks and its bottom are of cold water, whilst its current is of warm. There is in the world no other such majestic flow of waters. Its current is more

---

* See below, the chapter entitled, *The Earth and Man.*
† J. G. Kohl, *Geschichte des Golfstroms*, p. 1.

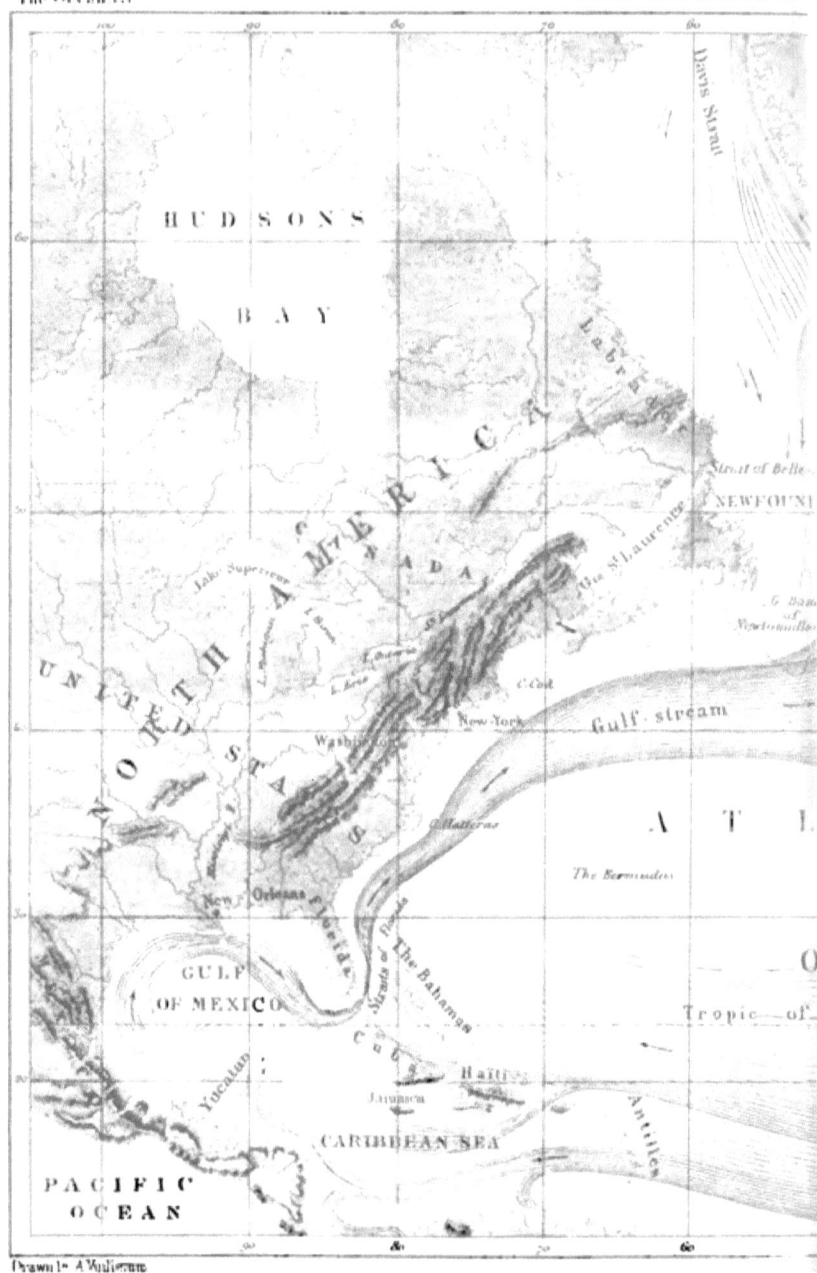

GULF STREAM

PL.V

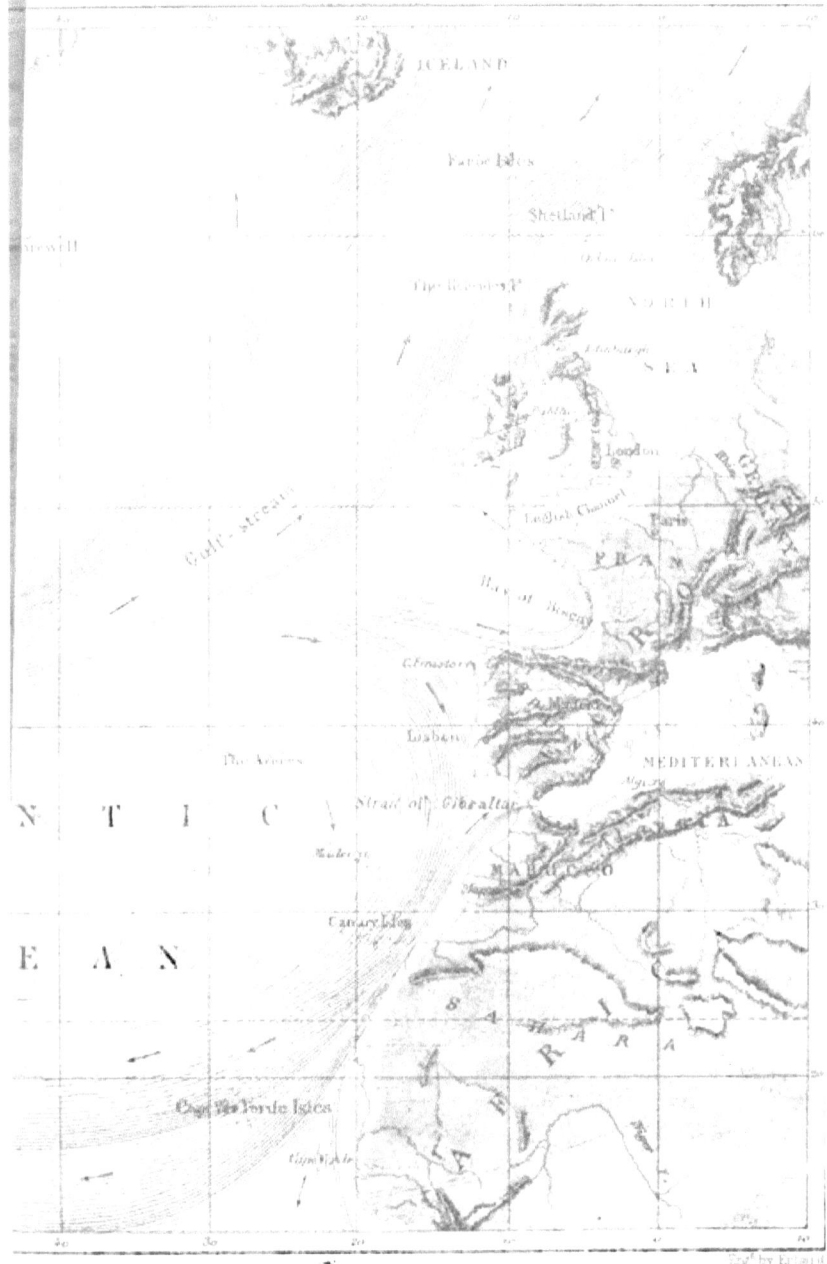

rapid than the Mississippi or the Amazon, and its volume a thousand times greater." Such is the epic language in which Maury's fine work commences.*

After having made the tour of the Caribbean Sea and the Gulf of Mexico in six months, after having driven back upon the shores of Alabama the muddy waters of the Mississippi which border its dark blue waves, the Gulf-stream follows the northern coasts of Cuba, then turns the southern point of Florida, and penetrates the strait which separates the American continent from the islands and banks of Bahama. Swelled by the mass of water which the great equatorial current sends directly through the straits of the archipelago, and above all by the old channel of Bahama, the Gulf-stream flows straight to the north, pressing through the ocean like a river nearly 37 miles wide, and of an average depth of 200 fathoms. Its speed is great, even equalling that of the principal rivers of the world, being sometimes from about 4½ to 5 miles an hour; but usually it is about 3½ miles. The mass of water discharged by the current may, therefore, be estimated at nearly 45 millions of cubic yards per second, that is to say, at 2000 times the mean discharge of the Mississippi; and yet it was to the outflow of this North American river that many geographers formerly attributed the existence of the Gulf-stream! When winds from the south, the west, or even the north-west, and the movement of the tides, favour the progress of this current, it rolls toward the Atlantic in much greater volume than usual. But on the other hand, when retarded by tempests that blow from the north-east, it pours a much smaller quantity of water into the ocean. When thus checked, it swells, rises, spreads with fury over the low lands that border it, ravages vast tracts, and causes whole islands to disappear. At its embouchure into the ocean, this marine river resembles those streams which flow through continents, it erodes on the one side, while it deposits alluvium on the other. And doubtless the Bahama islands, which are scattered through the sea to the east of the Gulf-stream, and the *Keys* or rocks developed on the north in a long range, rest on a foundation of submarine banks formed in part by the deposits of this grand river.†

On emerging from the strait of Bemini the Gulf-stream expands and spreads over the Atlantic, but at the same time its depth becomes proportionately less considerable. Whilst the strata of cold water

* *Physical Geography of the Sea*, p. 23.
† Agassiz. R. Thomassy, *Bulletin de la Société de Géographie*, Novembre, 1860. See below, the chapter entitled, *Earth and its Fauna*.

which serve as its banks retire on each side and allow it to spread over a greater breadth, the bed of cold water which bears it and

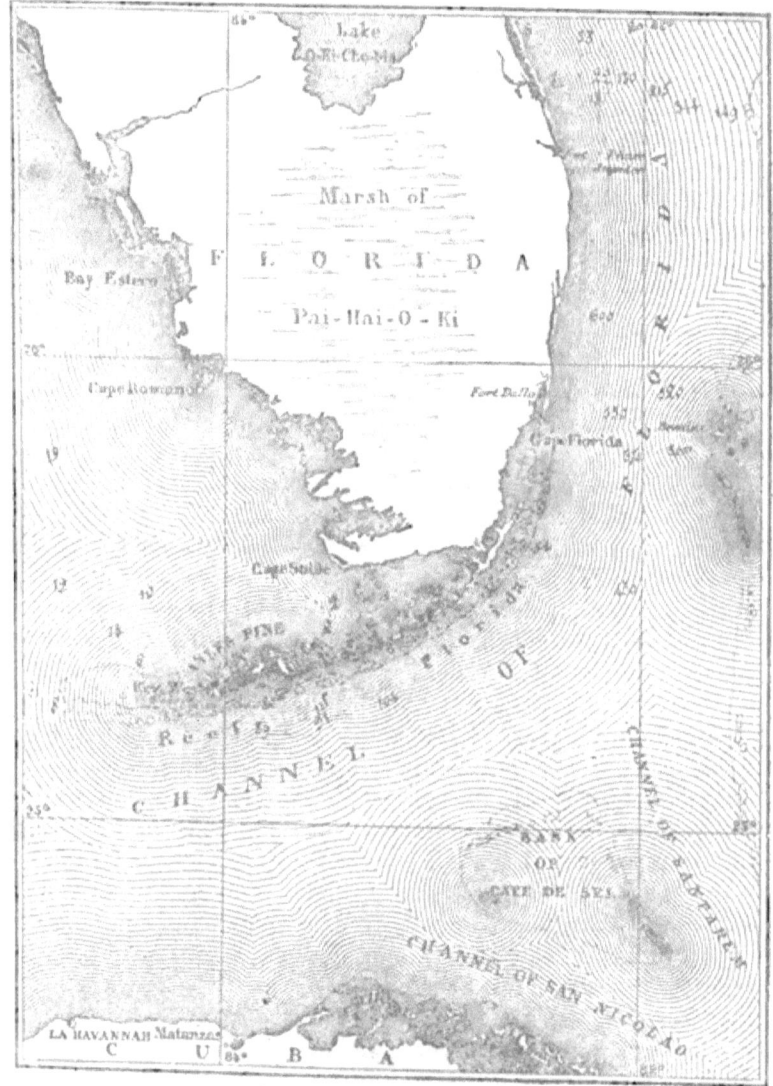

Fig. 22.—Channel of Florida.

over which it flows, as terrestrial rivers glide over beds of rocks, gradually approaches nearer the surface. At Cape Hatteras the depth of the current is about 120 fathoms, and its speed does not exceed three miles per hour; but it is twice as wide when it emerges from the strait of Bemini, and spreads over a space of about 78 miles. The thickness of this powerful stream of warm water is constantly diminishing, and when it has crossed the Atlantic it is only a superficial sheet. But even then, it covers an immense extent, reaching from the Azores to Iceland and Spitzbergen.

The soundings taken since 1845 by the Officers of the Coast Survey of North America, prove that the Gulf-stream flows along the coast of the United States at some distance from the land. The slight inclination of the low lands of Georgia and Carolina is continued under water till the lead reaches a depth of about 50 fathoms. The bottom then sinks rapidly, and forms a long valley parallel to the shore of the United States and the chalky walls of the Appalachians; it is in this valley, hollowed to the east of the submarine basement of America, that the waters of the Gulf-stream flow. Owing to the rotatory motion of the globe, and probably also to the general direction of the coasts, the current follows a constant direction to the north-east, and does not touch any of the advanced points of the continent. Off New York and Cape Cod it deviates more and more to the east, and ceasing to follow at a distance the American coast-line, rolls across the open Atlantic towards the shores of western Europe. Thus, as Maury says, if an enormous cannon had force enough to send a bullet from the strait of the Bahamas to the North Pole, the projectile would follow almost exactly the curve of the Gulf-stream; and gradually deviating on its way, would reach Europe from the west.

Between the 43rd and 47th degrees of north latitude, in the neighbourhood of the bank of Newfoundland, the Gulf-stream, coming from the south-west, meets on the surface of the sea the polar current discovered by Cabot in the year 1497. The line of demarcation between these two oceanic rivers is never absolutely constant, but varies with the seasons. In winter, that is to say from September to March, the cold current drives the Gulf-stream towards the south; for during this season all the circulatory system of the Atlantic, winds, rains, and currents, approach more nearly the southern hemisphere, above which the sun travels. In summer, that is to say from March to September, the Gulf-stream in its turn resumes its preponderance, and forces back the line of its conflict with the polar current more and more towards the north. Thus the great river undulates here and there over the seas, and according to the graceful expression of Maury, waves like a pennon in the breeze. But it is probable that the advance of the two opposing currents is often modified only in appearance, in consequence of the superficial expansion of cold or warm water. The bank of Newfoundland, that enormous plateau surrounded on all sides by abysses five to six miles deep, is undoubtedly due in great part to the meeting of these two moving liquid masses. On entering the tepid waters of the Gulf-stream the icebergs gradually melt and let fall the fragments of rock and loads of earth which they bear, into the sea. The bank which rises gradually from the bottom of the ocean is a sort of common moraine for the glaciers of Greenland and the polar archipelago.*

After encountering the waters of the Gulf-stream, those of the arctic current cease in great part to flow on the surface, and descend into the depths, in consequence of the greater weight which their low temperature gives them. The direction of this counter-current, exactly opposite to that of the Gulf-stream, is demonstrated by the icebergs which the warm breath of temperate latitudes has not yet melted, which travel towards the south-east, directly against the superficial current, which they divide like the prow of a ship. More to the south, we recognize the existence of this concealed current only by means of sounding apparatus, the cold waters serving as a bed to the warm river flowing from the Gulf of Mexico; it descends lower and lower as far as the straits of Bahama, where the thermometer discovers it at a depth of 220 fathoms.

Nevertheless a part of the waters of the polar current remains at the surface of the sea; and, gliding along the western coasts of the United States as far as the point of Florida, gives to the Gulf-stream by contrast very sharply defined limits. Generally the cold water coming from the Arctic Sea possesses sufficient force to compel the current from the gulf to bend sensibly towards the south, and to opppose an insurmountable barrier to it in the other direction. The warmest and most rapid part of the Gulf-stream, which forms precisely the left or western side of the current, is found in immediate juxta-position to a sheet of cold water, which spreads in an opposite direction between the Gulf-stream and the American shores. This counter-current, which interposes the waters of the Icy Sea between the coast of Carolina and the warm river flowing from the Gulf of Mexico, bounds

* See above, p. 16.

SALINITY OF THE GULF-STREAM.

the Gulf-stream like a wall of ice.* Sometimes the line of demarcation between the two liquid masses is so precise that it is appreciable to the sight, and the exact moment when a ship leaves one current, to cleave the other with its prow, may be distinguished. The water of the Gulf-stream is of a beautiful azure, that of the counter-current is greenish; the first is saturated with salt, the latter contains it in a much less proportion. The one is tepid, the other cold, and the thermometer, when plunged alternately in the two liquids, instantly marks the difference of temperature. On the boundary line of the currents, the friction of the two masses of water flowing in opposite directions produces a series of eddies, whirlpools, and short waves, which give to these ocean-rivers an aspect similar to that of continental rivers. Sometimes one can even hear like a dull roaring the noise of the waters contending on the surface of the sea. Floating plants and other fragments are whirled round on the ever-changing boundary of the two contending streams.†

The Gulf-stream, like all other currents, finally mingles with the sea, and thus tends to equalize the proportion of salt and all other substances contained in the liquid mass. The normal salinity of the Caribbean Sea is from 36 to 37 thousandths, except in the neighbourhood of the mouths of great rivers. After having received the fresh waters of the Mississippi and the visible and subterranean rivers of Florida, the Gulf-stream does not contain quite 36 thousandths of saline substances; but this is gradually increased as it advances towards the north. Off Newfoundland, where the waters of the St. Lawrence and many other rivers, as well as the melted ice, fogs, and heavy rains, have rendered the waves of the sea more fresh, the Gulf-stream contains less than 34 parts in 1000 of saline matter, but it gradually increases the proportion to 35 thousandths as it shapes its course towards the coasts of western Europe, and the polar regions. The currents of cold water which serve as its bed are all less rich in saline substances, as Forchhammer and other chemists have proved. But in consequence of the incessant mixture of the waters, an equalization of saltness between the currents is produced in the various latitudes.‡

Another effect of the Gulf-stream, no less important in the economy of our planet, is that which it accomplishes in concert with

---

* Franklin Bache, *United States Coast-Survey.*
† Kohl. Fitzroy, *Adventure and Beagle,* Appendix to Vol. II.
‡ Forchhammer, *Philosophical Transactions,* part I., p. 241, 1865.

the south-west winds,* on the climate of western Europe. While rotating in the Gulf of Mexico, as in an immense cauldron, the waters of the current are gradually heated; when they escape through the strait of Bemini to enter the ocean, their temperature is not less than 86° Fahr., and exceeds by about 4° Fahr. the natural heat of the neighbouring beds of water. The waters of the Gulf-stream lose their warmth but slowly, and during winter they often have, off Cape Hatteras and the bank of Newfoundland, a temperature exceeding by 21° or 28° Fahr. that of the rest of the Atlantic under the same latitudes. When the Gulf-stream meets the polar current, the former has still a temperature of 36° or even 45° Fahr., whilst, even at a distance of some hundreds of miles from the coasts of Labrador, the latter is sometimes found to be below freezing-point (24·8° Fahr.); thus, in defiance of latitude, the waters of the tropics and of the icy zone are brought into juxtaposition. In its advance towards the north, the upper strata, which in consequence of radiation have become colder than the subjacent layers, descend to a greater or less depth in the mass of the current, and are replaced by the warmer and lighter water, lying immediately below. Thus a constant alternation of position is produced in the liquid strata of the Gulf-stream, and one may remark in consequence, in crossing the whole breadth of the current, a series of parallel bands of unequal temperature.† In each of these bands the warm water rises by turns to the chilled surface of the sea. It is a remarkable fact that if the Gulf-stream did not flow as it does in a bed entirely composed of cold water, but moved along the very bottom of the ocean, it would rapidly lose its high temperature, and would cease in consequence to be a source of heat for western Europe. In fact, the earth being a better conductor of heat than the water, the warm waters of the current would communicate their temperature to it, and would finally lose their whole store. But the cold waters of the polar current, being interposed between the bottom of the sea and the waters of the Gulf-stream, serve as a protecting screen to the latter and hinder their refrigeration. It is by such contrasts as these that the harmony of the world is established.

The quantity of heat which the Gulf-stream carries towards the northern regions forms a very considerable part of the caloric stored up in its waters under the tropics. The cetaceans, fish, and other inhabitants of the torrid zone follow the course of the Gulf-stream without

---

* See the chapter entitled, *Climates*.
† Franklin Bache, *United States Coast-Survey*.

perceiving that they have changed their country, and often push their adventurous voyages to the Azores and even to the coasts of Iceland; the southern birds mount also towards the north in the warm stream of air reposing on the current. The animals of northern seas, on the contrary, are kept prisoners in the glacial ocean, and the right whales, says Maury, recoil before the Gulf-stream, as before "a barrier of flame." The total warmth of the current would suffice, if it was centred in a single point, to fuse mountains of iron and cause a river of metal as mighty as the Mississippi to flow forth. It would suffice also to raise from a winter, to a constant summer temperature, the entire column of air which rests on France and the British Isles. But, though it is spread over enormous spaces to the west and north of Europe, the Gulf-stream does nevertheless exercise a preponderating influence upon the climate of this part of the Old World. Owing to the warmth of its waters the lakes of the Färoe and Shetland Isles never freeze during winter; Great Britain is enveloped in fogs, as in an immense vapour-bath, and the myrtle grows on the shores of Ireland, the "emerald isle of the seas," under the same latitude as Labrador, that land of snow and ice. In green Erin, an island privileged in so many respects, the western coasts (the first land which the Gulf-stream encounters after crossing the Atlantic) enjoy a temperature two degrees higher than that of the eastern coasts. In spite of the path of the sun, it is on an average as warm in Ireland under the 52nd degree of latitude, as in the United States under the 38th degree, or about 1025 miles nearer the equator.

The Gulf-stream, which conveys the tropical warmth to the temperate countries of Europe, very often serves as a high-road for tempests. Hence the names of *weather-breeder* and *storm-king*, which have been given to this current.* The movements of the atmospheric ocean and those of the ocean of waters occur in such complete parallelism that one would be tempted to regard them as one and the same phenomenon in the *ensemble* of aërial and marine currents.† Thus the Gulf-stream seems to be for the winds as it really is for the waters, the great medium between the old and new worlds. It carries to the seas at the north of Europe the salinity of the Gulf of Mexico; it bears with it the warmth of the tropics for the advantage of the temperate regions, and marks the track which the torrents of electricity, disengaged by the hurricanes of the Antilles, follow. It is indeed the great serpent of the Scandinavian poets, which uncoils its immense

* F. Maury, *Geography of the Sea*.
† See below, the chapter entitled, *The Air and the Winds*.

folds across the ocean, and from its head, which it waves here and there over the shores, wafts a gentle breeze, or pours forth storm and lightning.

The Gulf-stream crosses the Atlantic with a mean speed of about 24 miles a day, as has been ascertained either by direct measurement at different parts of the ocean, or by means of notes, which having been thrown overboard in bottles carefully closed, have floated for weeks or months at the will of the waves, and then been fished up in other latitudes or found on some sea-shore. In their long passage, the deep waters of the marine river of America transport scarcely any other alluvium than the living frustules of animalcula, which fill the tepid waters of the current and are constantly falling in a kind of snow to the bottom of the sea. But here and there on the surface of the Gulf-stream float trunks and branches of trees, which are finally thrown on some coast of Europe, and even on the island of Spitzbergen. It was these remains which our ancestors of the middle ages believed to come from the fabulous island of St Brandan or from Antilia, and which furnished matter for thought to daring navigators like the great Columbus.* Seeds carried from the New World by the current have found a favourable soil on the shore of the Azores, and although many thousands of miles from their native land, have germinated and borne fruit. Often too the waves of the Gulf-stream bring to Europe the broken products of human industry and the timber of wrecked ships. During the Seven Years' War, the mainmast of an English ship-of-war, the *Tilbury*, which had been burnt near St. Domingo, was found on the northern coasts of Scotland. In the same way a river-boat, laden with mahogany, was once even driven to the Färoe islands. The remains of ships, wrecked in the latitude of Guinea, have been brought to the coast of the British Islands, after having twice crossed the ocean in opposite directions; and Esquimaux have often been carried by the waves to the Orkneys.†

It is rather difficult to lay down the precise route of the Gulf-stream in the seas of western Europe, because of the enormous width of its moving expanse. One may say that in reality it stretches over the whole ocean, from the Azores to Spitzbergen; but having lost in its onward impulse in proportion as it has gained in extent, it is modified and turned aside in its course by a host of local circumstances and the varied configuration of the coasts of Europe. Only that part of the cur-

* F. G. Kohl, *Geschichte des Golfstroms*, p. 17.
† Humboldt, *Ansichten der Natur*, notes.

rent which flows to the north of Ireland and Great Britain maintains its original direction. It bathes all the islands between Scotland and Iceland, warms the coasts of Norway, even in Lapland it melts the ice at the port of Hammerfast, and then continues its course in the Polar Sea towards Spitzbergen. Thus, as the Swedish expedition in 1861 ascertained, the current makes itself felt even on the northern shores of the latter archipelago; for the seeds of a plant from the Antilles (*Entada gigalobium*) were found on the shore of Shoal-Point, lying at more than 80 degrees north latitude. Indeed it is certain that the current even bathes the western coasts of Nova-Zembla, for bottles that came from a glass factory at Norway, and the nets of Scandinavian fishermen, have been found there.

How then do these waters, which spread in such a vast sheet over the surface of the Icy Sea, continue their progress towards the Pole? Here hypothesis commences, since no navigator has yet been able to explore these latitudes and study their hydrological laws. But we know at least in part the origin of the polar current, and by the direction which this mass of water takes, may be indicated that which the Gulf-stream itself must follow. Along all the northern coasts of Siberia, as Wrangel and other explorers have told us, a current of cold water flows from east to west. Encountering on its way the large island of Nova-Zembla, it covers the strand and rocks with enormous quantities of ice, which render the island quite uninhabitable and close the straits to navigation. Arrested by this barrier the waters of the glacial current are forced to bend to the north, and flow in a north-westerly direction towards Spitzbergen, round the northern archipelago of which they finally turn in order to enter the seas around Greenland. It is here that they begin to take a direct road towards the equatorial seas; and all the navigators who have ventured to the north-west of Iceland have recognized the existence of this stream, flowing along the coast-line as far as Cape Farewell.. Its average speed, according to Graah and Scoresby, is from 3 to 4 miles a day.

To the south of Greenland the lessened sheet of the Gulf-stream must meet this transverse current, and doubtless, in consequence of the greater weight which its stronger proportion of saline substances imparts to it, it plunges into the depths and is changed into a submarine current, which finishes by mixing completely with the cold waters of the northern seas, and flows back at last towards the equator in an opposite direction to that which it at first pursued. Thus the river of warm water from the Gulf of Mexico feeds by its incessant contributions the polar counter-current, and the great circuit is established between the zone of heat and that of ice. Perhaps, too, the reflux of the Gulf-stream is partially accomplished, under the pressure of water from the north by an abrupt turn. This would explain the strong salinity of 35 thousandths, which Forchhammer found in the waters of the polar current to the east of Greenland.

It is not only in the wide extent of the North Atlantic, from Nova-Zembla to Iceland, that the Gulf-stream takes a submarine course; the same is the case, it appears, in Baffin's Bay to the west of Greenland. In fact, from Cape Farewell to eight degrees further north the existence of a coast-current has been ascertained, which carries the ice in an exactly contrary direction to that of the current, which follows on the west coasts of Labrador, and which serves as a high road for the fragments of the ice-fields.* This current was formerly considered as the continuation of the one which flows along the eastern coast of Greenland from north to south, and which would thus have abruptly turned round Cape Farewell. But it is much more natural to think that the polar current continues its route directly towards the great centre in the tropical seas. In this case, the current on the western coast of Greenland would be simply a branch of the Gulf-stream; which is rendered almost certain by its waters being comparatively warm. The sea very seldom freezes on the shore which it bathes, and the climate there is on an average nine degrees (Fahr.) warmer than on the coast looking towards the east. Towards the 78th degree of latitude, this river-like current completely ceases, and it is undoubtedly there that it becomes submarine, perhaps to flow on the surface again in the open sea of Kane.†

On the other hand, if in the icy seas the various branches of the Gulf-stream change into smaller counter-currents, the polar currents do the same more to the south, and become the bed for the waters which flow to the north. These contain, it is true, more saline substances, but they are also warmer; rendered heavy by the salt, they are lightened by their high temperature, so that a slight difference of warmth or of salinity can modify their equilibrium and make them change their position with the polar current. In the temperate seas, where they are still warm and strongly saline, they flow on the surface; but sink on the contrary in the icy seas, where they are chilled or where the admixture of salter water is effected. This explains the intersection of the currents. To the north of Spitzbergen and Nova-Zembla, the

* See above, p. 43.
† Graah.—Mühry, *Mittheilungen von Petermann*, t. ii. 1867.

Gulf-stream is a submarine sheet; to the south of Iceland it is the waters from the Pole which flow below. Not far from the Färoe islands the sounding-lead can even indicate the direction followed by the icy counter-current, owing to the layer of volcanic remains, which have been brought from the coasts of Iceland, and spread over a space of 25 degrees of longitude between the 47th and 52nd degree north latitude. This hidden river must flow, at least in certain places, on the very bottom of the sea, for various soundings taken by McClintock to the south-east of Iceland, show that all the light detritus has been swept away by its waters.*

If the Gulf-stream throws out various branches towards the north, which contribute to form the vast circumpolar whirlpool in the same way, another branch flowing towards the south goes to swell the equatorial current. This offshoot of the Gulf-stream, of which one branch penetrates into the Bay of Biscay and forms the coast-current called Rennell's,† flows along the coasts of the Iberian peninsula, follows the outline of Africa to the south of the Canaries and Cape de Verde islands, where lateral counter-currents occur, and enters the great marine river which moves the waters from east to west, "following the course of the heavens."

Thus is completed the immense circuit of the Atlantic, in the centre of which the sea meadows of wrack † extends in clusters like an archipelago. It is owing to this perpetual circuit that navigators in sailing vessels have been able to reach the New World from western Europe. If Columbus had not made use of the semi-circular current which flows from the coasts of Spain to the Antilles, he certainly would not have discovered America. If the pilot Alaminos, and, since his first voyage, the greater part of the navigators returning from the Antilles and United States, had not, either without knowing it, or else well understanding the cause, followed the course of the Gulf-stream, the coasts of America would have remained practically far more distant from Europe than they really are. The colonies, now become so prosperous as independent republics, would be still in deplorable isolation; and civilization would have been greatly retarded, or even completely arrested, for want of new impetus. As to commerce, properly so called, one can judge of the influence exercised upon it by the movement of the waters of the Atlantic, when one examines on a map the position of the great centres of trade. Havannah and New Orleans, two

---
* Wallich, *North Atlantic Sea-bed*.
† See below, chapter entitled, *Earth and its Flora*.

principal markets of the Antilles and Mississippi States, are, so to say, at the source of the Gulf-stream. New York is situated facing the principal bend of this current, at the spot where the vast river flowing from the Antilles bends towards Europe. Finally, Liverpool, among so many other considerable ports washed by the Gulf-stream on its arrival at the coasts of the Old World, is the one which is most directly in the path of its waters.

When Franklin discovered, in 1775, that the mariner has only to plunge a thermometer in the water of the Atlantic to discover if he is sailing over the Gulf-stream or outside its course, the illustrious savant immediately perceived the importance of this fact for navigation. He even thought for a long time he must conceal it, from a fear that the English government, then at war with the American Colonies, would profit by this discovery to send ships and men more rapidly against the revolted provinces. After the definite establishment of American Independence, no peril of this kind being any longer to be feared, all navigators were enabled for the future to know precisely the high road which they had to follow in the open sea to reach Europe most expeditiously from America, and what particular line to avoid in order to effect the journey in an opposite direction. Towards the middle of the last century, the whalers

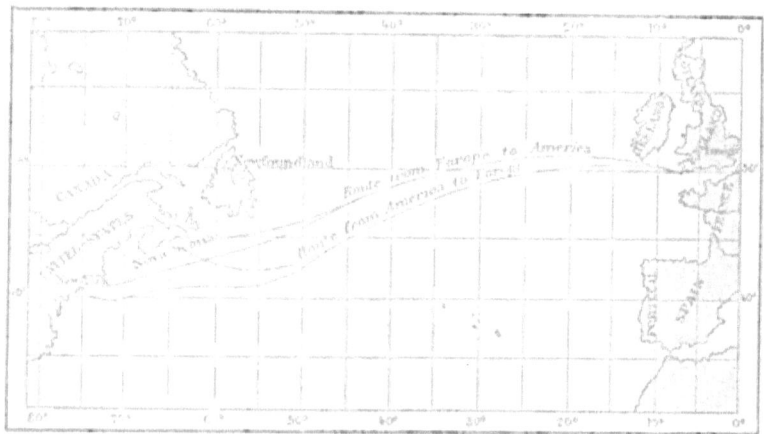

Fig. 23.—Route of Steam-packets, after Maury.

of Nantucket and the skippers of Rhode Island had already from experience come to choose two different **routes for going and returning.**

In order to "descend" on England they allowed themselves to be carried with the Gulf-stream, and on their return crossed this current at the banks of Newfoundland, and "mounted" the Arctic countercurrent;[*] on these voyages they distanced vessels from other seaports, on an average, by 74 miles per day. The progress of navigation permits us now to utilize the impelling force of the currents of the North Atlantic far better than the sailors of Providence could. The normal duration of the passage has been reduced to half. Eight weeks were formerly reckoned for a voyage from England to the United States; now four weeks suffice for sailing vessels, and some have even made the journey in seventeen days only. Steamers, which also have a double route too, in order to avail themselves of the current, accomplish the passage in nine or ten days. For commerce, civilization, and the brotherhood of peoples, such a result is not less important than as if the continents themselves were shifted, so as to reduce by three quarters the width of the ocean which separates them.

[*] J. G. Kohl, *Geschichte des Golfstroms*, p. 103.

CHAPTER IX.

CURRENTS OF THE SOUTH ATLANTIC AND THE INDIAN OCEAN.—DOUBLE EDDY OF THE PACIFIC OCEAN.

THE circuit of the waters which occurs to the south of the equator, in the southern basin of the Atlantic, is much less known than that of which the Gulf-stream forms a part, but all that has been observed of it by navigators proves that the movements of the liquid mass are analogous in the two hemispheres. A current of cold water, coming from the Antarctic seas, dashes against the Lagullas Bank, to the south of the African continent, and divides into two branches, one of which re-enters the Indian Ocean, while the other flows along the western coast of Africa, penetrates into the Gulf of Guinea, and, in consequence of the motion of the earth, bends towards the west in a wide semi-circle. To the south of the Cape Verde Islands, the waters coming from the southern seas join those which flow from the Icy Sea of the north, and uniting into one river of 500 to 1000 miles wide, move slowly in the direction of South America and the Antilles. The greater mass of water approaches the continent to the north of Cape St. Roque, the advanced promontory of Brazil, and flowing to the north-west along the coasts of Guiana and Columbia, enters the Caribbean Sea, there to form the Gulf-stream. A less considerable fraction of the equatorial current bends to the south of Cape St. Roque, and follows the Brazilian coast-line to the south-west. But in descending towards latitudes nearer and nearer the southern pole, the warmer current from the equator incessantly gains on the rotatory movement of the earth; consequently, it bends more to the south than to the south-east, and forming a sort of gulf-stream in an opposite direction, it strikes the polar current to the east of the Falkland Isles, whose position in the southern hemisphere corresponds to that of Newfoundland in the northern hemisphere. There the warm current, after having deposited drift-wood, taken from the Brazilian coast, on the shores of the Falkland Islands, sinks below the lighter strata of the glacial current; while the latter directs its course to the north-east towards St. Helena, where it joins the great equatorial river. The whole circuit is accomplished in a period which may be estimated at about two or three years.*

Dissimilar, and often contradictory, observations recorded by various navigators who have studied the phenomena of the waters in the South Atlantic, seem to put it beyond doubt that the currents of this basin have not the same regularity of course as those of the Northern Atlantic. It frequently happens that the water does not flow in the direction indicated on maps, or even tends in an opposite direction to the normal movement. The reason of this difference between the two basins is quite natural. While the North Atlantic is a very regular sea in its general form, bounded on each side by almost parallel shores, the marine area lying between Africa and South America expands very widely from the coast of the southern polar land. It may be considered simply as a gulf of the great ocean, which extends around the globe to the south of the extremities of the three southern continents. As a consequence of this irregular disposition of the coasts, the variations from the normal circumstances of the waters cannot fail to be very great. The cold waters from the Antarctic Pole, charged with fragments of ice-fields and icebergs, flow it is true with a continuous motion to replace the vapours which rise incessantly from the equatorial Atlantic. But the regular play of the currents is modified now at one point now at another, according to the greater or less activity of evaporation in those parts of the sea. Besides, the changing coast-winds which blow alternately from the ocean to the land and from the land to the ocean, impress their varying movements on its surface.

The Indian Ocean has likewise its great circuit of water. There too the mass of fluid, chilled by its sojourn in the icy zone, is incessantly flowing towards the equator, in order to fill up the vacancy produced by the annual evaporation of 13 to 16 feet. It flows along the western coast of Australia, and afterwards unites with the waters that come from the Pacific Ocean, through Torres Straits and the East Indian archipelago. But there the regular current seems to lose itself; and we only see in the Gulfs of Bengal and Oman, marine rivers changing their course with the monsoons. Nevertheless it must really be that the general movement of the waters is continued from the east to the west around the vast basin, for on the eastern coast of Africa a current of warm water, incessantly supplied by the seas which bathe Hindostan and Arabia,

* *Mittheilungen von Petermann*, t. x. 1866.

flows to the south-west, and under the name of the Mozambique current, passes between the island of Madagascar and the continent, touches the edge of the great submarine bank of Lagullas, and spreads into the Antarctic Ocean, after having mingled a part of its waters in the great whirlpool of the Atlantic. At the part where it is narrowest, the Mozambique current is almost as rapid as the Gulfstream, and moves with a speed of about 4½ miles an hour. In the centre of the eddy in the waters of the Indian Ocean, as in the North Atlantic, whole meadows of seaweed spread over the calm waters.

The circuit of the currents commences in the great Pacific Ocean in the same manner as in the other basins. An immense river of cold water of unknown breadth strikes the island of Magellan, at the south of America, and divides into two partial currents, one of which, penetrating into the Atlantic to the east of the Falkland Isles where ice never comes, joins in the great round of waters between Africa and Brazil, while the other flows directly to the north along the coasts of Patagonia, Chili, and Peru; this is Humboldt's current, thus named after the celebrated traveller who first recognized its existence. It carries with it large icebergs often laden with stones and fragments that have fallen from the Antarctic mountains, and by the coldness of its waters produces a remarkable lowering of the temperature in all the countries whose shores it bathes. This liquid mass, which has a depth of no less than 670 fathoms on the coast of Chili, gives to the vegetation of the country a remarkable analogy with that of St. Helena, which at a distance of 4000 miles is washed by another branch of the Antarctic current. Humboldt and Duperrey state, that off the coasts of Callao and Guayaquil, that is to say, in one of the driest climates and most exposed to the rays of the sun, the current is on an average at from 59° to 60° Fahr., while the adjacent seas are about 20° warmer. Not a branch of coral can take root on the rocks and shores washed by this current of cold water: the polar current changes everything on its passage, the flora, fauna, climate, and even the history of mankind. If the air was not constantly refreshed by the contact of cold water coming from the Pole, Peru, which is so rarely watered by rain, would be transformed into another desert of Sahara, and human life would become almost impossible there. By this current, too, the distances are notably diminished, and Valparaiso, Coquimbo, Arica, Callao are, in reality, less distant from Europe than they appear on the map; for after having rounded Cape Horn, the ships sailing along the western coasts of South America, are carried about 15 to 20 miles a day by this current.

Widening more and more on the side of the open sea, Humboldt's current ends by abandoning the coast-line, and bending towards the west, to mix its waters with those of the equatorial current tending from east to west across the Pacific. This liquid moving mass is undoubtedly the most powerful oceanic river of our planet. According to Duperrey, it has a mean width of no less than 3500 miles, from the twenty-sixth degree of south latitude to the twenty-fourth degree of north latitude, and on its immense journey in a straight line round the world, it traverses from 130 to 140 degrees of longitude; that is to say, more than a third of the circumference of the globe. Its average speed is like that of Humboldt's current, about 19 miles per day, but in certain places according to the seasons an advance twice as rapid has been ascertained. What the quantity of this enormous mass of water can be that is thus displaced from one end of the sea to the other, is unknown; for it would be first necessary to know the mean depth of the current, but this the sounding-lead has not yet discovered. It is only known that at the point where the water from the pole turns towards the west, to enter the great equatorial stream, it proceeds "en masse" in one direction, with a depth of not less than a mile.

In the midst of the innumerable islands which are scattered over the Pacific, the general regularity of the great current is frequently disturbed, at least on the surface, in consequence of evaporation, rains, and even by the incessant labours of the coral-building zoophytes, which in various ways disturb the equilibrium of the ocean. But under the threefold influence of the terrestrial rotation, the trade-winds, and the great tidal wave which is propagated from east to west across the ocean,* the quantity of water moved each day towards the west is certainly several tens of thousands of cubic miles. The only anomaly in this prodigious movement of the waters of the Pacific which seems inexplicable is the existence of an oceanic river flowing in an opposite direction to the principal current. This reflux has been observed to the north of the equator over a mean breadth of upwards of 300 miles; elsewhere its speed is variable, and its advance is not always in the direction of due east. In the absence of measurements and positive experiments which permit us to give an exact account of the progress of this counter-current in the different seasons, several hypotheses have been suggested to explain its origin. The common opinion is that it is masses of water turned aside on their course and thrown back by submarine plateaux.†

---

* See below, p. 101.   † Herschel, *Physical Geography*.

Nevertheless it is much simpler to admit that this is a normal phenomenon, for in the Atlantic Ocean it has also been established that some lateral eddies tend in an opposite direction from the great liquid mass flowing from east to west.

When it has arrived at the end of its voyage across the Pacific, the equatorial current must of necessity change its direction. A portion of its waters, driven now in one direction and now in another by the monsoons which succeed one another on the borders of the continents of Asia and Australia, flows into the Indian Ocean by the shallow straits of the East India islands. But the greater mass of the current is thrown back either to the south or to the north, by the resistance of the shores against which it dashes and breaks. The half of the current which strikes the coasts of Australia diverges towards the south, and flows in the direction of the Antarctic lands. It thus flows in the opposite direction to the polar current, which it finally encounters to the south of New Zealand, and plunges beneath its colder waters which by their freshness are rendered lighter. To the east and north-east the current from the Antarctic seas completes the enormous circuit described by the waters in the southern basin of the Pacific.

The other half of the equatorial current, diverted by New Guinea, the Philippines, and that long barrier of islands lying to the east of China, bends gradually towards the north and flows along the outer coasts of Japan. It is the Gulf-stream of the Pacific Ocean, called also Tessan's current, after the navigator who revealed its existence to the savants of Europe. But for centuries, and perhaps thousands of years, the Japanese have known and prized it highly for their coast-navigation. They give it the name of Kuro-Sivo, or "Black River," doubtless because of the deep blue of its waters. Less rapid than the Gulf-stream, its advance is nevertheless on an average more than 1½ mile per hour, and in many places very much exceeds this speed. Before Yeddo its mean temperature is $75.2°$ Fahr., that is to say, about $10°$ to $12°$ Fahr. higher than the still waters beside it. Furthermore, the Kuro-Sivo, like the Gulf-stream, is composed of liquid bands of unequal temperature flowing beside each other like two distinct rivers in the same bed.

In passing the largest island of Japan,[*] the Black River, obedient to the impelling force which the rotation of the earth has communicated to it under tropical latitudes, already commences to bend towards the north-east, and, spreading over a vast extent, loses in depth what it gains in surface. To the north of Japan, it meets

---

[*] De Kerhallet, *Considérations sur l'Océan Pacifique*.

obliquely a current of cold water emerging from the Sea of Okhotsk, to replace in part the void caused by the evaporation in the equatorial seas. Thick fogs, similar to those of the Banks of Newfoundland, rest above the spot of contact between the warm and cold waters. Shoals of fish, the object of pursuit to fishermen, people this maritime zone, which serves as a limit between the two currents, and where the mass of animalcula and remains brought from the tropics is joined to those which are conveyed in the waves coming from the north. Still, the phenomena presented by the meeting of the two currents have not the same grandeur in the North Pacific, as under the corresponding latitudes of the Atlantic. For the mass of water flowing from the Sea of Okhotsk is relatively less considerable, and the opening of Behring's Straits, 31 miles wide and 50 fathoms deep, is of too small dimensions to allow much water from the icy ocean to penetrate into the Pacific. Only small coast-currents carrying the pines and firs from the shores of Siberia, and rounded ice-floes from along the two coasts, cross from one sea to the other. In summer the current which comes from the north, both on the eastern and western bank of the strait, is only a superficial current. On the other hand, the slight portion of the waters of the Black River which passes beyond the range of the Aleutian Islands, to enter Behring's Straits, is a submarine current, at least during the summer season. Arriving in the icy sea, still warm and strongly saline, it mingles with the cold and light water which descends into the Atlantic by Baffin's Bay.*

The great mass of the Kuro-Sivo traverses the Northern Pacific from east to west with a graceful curve, no less beautiful than that formed by the islands that are washed by its waters; then bends gradually to the south-west and south, to coast the shores of California; finally, in the neighbourhood of the tropics, it changes its direction again, and is lost in the equatorial current, enclosing in its circuit a floating forest of sea-weed hardly less extensive than that of the Pacific.

Contrary to Humboldt's current, which rolls its cold waters and drives before it icebergs to refresh the dry and burning atmosphere of Peru, the gulf-stream of the Japanese carries along the coasts of Sitka and Vancouver's Island a mass of waters warmed by a long sojourn under tropical heat, and by its vapours brings spring to regions which without it would have a very severe winter. It bears on its waves the fragments which it has received from the coasts of

* De Haven—Mühry—Gustave Lambert.

the Moluccas, the Philippines, and Japan. To the inhabitants of the Aleutian islands and Alaska it gives, as fuel, camphor-wood, and other odoriferous trees from southern countries; it serves too as a highway for all kinds of waifs, carries away disabled ships, and numberless traditions relate how Japanese sailors were drifted afar and landed against their will on the coasts of America. And it is perhaps to an adventure of this kind that the Chinese navigators owe their discovery of the New World ten centuries before Columbus, if it is true that the country of Fusang, mentioned in the annals of China, is in fact the countries of Mexico and Guatemala. Messrs. Neumann, d'Eichthal, and other learned scholars do not doubt the authenticity of this historical fact.

CHAPTER X.

LATERAL EDDIES.—RENNELL'S CURRENT.—COUNTER-CURRENT IN THE SEA OF THE ANTILLES.—EQUILIBRIUM OF THE WATERS IN THE BALTIC, THE BOSPHORUS, AT THE ENTRANCES TO THE MEDITERRANEAN AND THE RED SEA.—EXCHANGE OF WATER AND SALT BETWEEN THE SEAS.

NONE of those great currents, which wind through the oceanic basins, show in their exterior contours the same sinuosities as the seas through which they flow. While most of the shores present a succession of promontories and gulfs, the currents stretch in long regular curves, which in their vast sweep indicate but generally the form of the depression which contains them. Every considerable gulf, which is separated from the ocean by any projecting land, remains outside the whirlpool of waters, unless it should be in the very axis of the current, like the sea of the Antilles. Yet even in those parts, which do not share in the general circulation, the waters do not remain perfectly stationary. They also have their circulatory system, and it is from the great maritime current that each secondary eddy receives its impulsion.

A remarkable example of these currents of the secondary order is presented on the west of Europe in the semicircular basin formed by the coasts of Spain, France, England, and Ireland. A portion of the waters of the Gulf-stream coming from the north and north-west strikes the coasts of Galicia and the Asturias; it turns east towards the extremity of the Gulf of Gascony, flows along the shore of the Landes, then that of Saintonge, Poitou, Bretagne, and returning in a north-west and west direction, forms a sort of liquid barrier across the Channel. To the south of Cape Clear this oceanic river, known under the name of Rennell's current, after the English savant who discovered its existence, finally enters the Gulf-stream, and returns to the south with the waters of the ocean. Thus a complete circuit is made around the basin, analogous to that which occurs in each of the great oceans of the world. Rennell's current, in its turn, coasting at a greater or less distance the shores of the continent, sends out into the little bays currents of a third order, which also complete their circular movement, like the Gulf-stream and the Kuro-Sivo; and so by lateral transmission, the circulation of the waters is continued from oceans to gulfs, from gulfs to bays, and from these to the creeks. These secondary currents, however, are usually much less regular than the general currents, and navigators have ascertained that at times Rennell's current has flowed in a completely reversed way to its normal direction.*

Secondary currents generally move in a course exactly opposite to that of the principal stream, of which they are only a branch bent back on itself. Either permanently or temporarily they are found in all seas, open or inland, in all gulfs and bays of the ocean. Even the sea of the Antilles, the waters of which are carried almost *en masse* towards the Gulf of Mexico, presents at its western extremity a permanent current, which tends from the shores of the isthmus to those of Columbia. A vessel drifted by the principal current into the neighbourhood of Nicaragua, would only have to ascend to Colon, and then abandon itself to the waves in order easily to accomplish its return voyage, borne by the waters which flow incessantly in the direction of Carthagena and Santa Marta. Many lazy seamen never pass from the ports of the isthmus to those of Terra Firma in any other way. Regardless of time, they let themselves be rocked by the billows without even taking the trouble to hoist the sails. Their bark, slower than a tortoise, advances at the most but a mile an hour, and after eight or ten days spent on the passage, they finally perceive the bluish mountains of New Granada, and its sandy shores shaded by cocoa-nut trees.

There are some currents which are evidently produced by a disturbance of equilibrium between two levels. Thus the Baltic Sea, which receives more water by the contributions of rivers than it loses by evaporation, must necessarily distribute its superfluity in the North Sea through the straits of the Sound and the two Belts. Nevertheless, these outlets being large and deep enough to diffuse the superabundant water in a little time, the current is not permanent. Waves from the North Sea, driven into the Baltic by the westerly winds, frequently meet it, and from this conflict of waters arise local and unexpected movements, dangerous to ships. Every four days the waters on the surface flow on an average for forty-eight hours towards the Cattegat, then flow back into the Baltic for

* Gareis and Becker, *Physiographie des Meeres*.

one day, and during another day there is no sensible movement in either direction. Often, too, according to Forchhammer, the two contrary currents glide one above the other; the lighter on the surface, coming from the Baltic, and the other from the North Sea, heavier by reason of the salt it contains, flowing beneath.

At the other extremity of Europe similar phenomena occur in the Bosphorus, at the outlet of the Black Sea. This strait, which receives the superabundant waters of the Euxine, presents a mean breadth of more than a mile, with a depth of 15 fathoms,* so that if the waters of the sea flowed there in a continuous manner as in the bed of a river, and the swiftness of the current were only about $1\frac{1}{4}$ mile per hour, it would not discharge less than nearly 36,000 cubic yards per second. But it is probable that all the united affluents of the Black Sea and the Sea of Azof supply hardly the half of this quantity; and, besides, a great part of the water brought by them is carried off again by evaporation. The Bosphorus is therefore too large to serve as the bed of a single current flowing from the Black Sea into the Sea of Marmora. It has been observed that the waters ordinarily descend towards the Mediterranean, with a speed of from 2 to 4 miles an hour; but the existence of tolerably rapid lateral counter-currents has also been established; and sometimes the winds blowing from the west cause the principal current to flow back through the strait. A submarine movement of the waters in the direction of the Black Sea also exists, as already ascertained by Marsigli in the last century.

At the western part of the Mediterranean, between Gibraltar and Ceuta, the normal current is that coming from the ocean. In fact, the Mediterranean has not many considerable tributaries. It only receives a single river having a really great mass of water, namely, the Danube; its other affluents of any importance, the Rhone, the Po, the Dniestr, the Dnieper, the Don, and the Nile, bring, on an average, not more than 19,620 cubic yards of water per second.† On the other hand, evaporation is very rapid in the basin of the Mediterranean, especially on the coasts of Egypt and Tripoli. We may admit that the quantity of water taken from this basin by the solar rays, and not directly restored by rain, annually represents a section of about $4\frac{1}{2}$ feet; which is probably near enough to the truth, as in the neighbourhood of Genoa, Beaucaire, Arles, and Perpignan, on the northern shores of the sea, the evaporation exceeds four-tenths of an inch per day in the great heat, and nearly 2 feet

* Tchihatchef, *Asie Mineure*.
† See in Vol. I. the section entitled, *Rivers*.

during the three summer months,* while the amount of rain during the year is about 19½ inches. The result is that the Mediterranean constantly loses three times as much water as it receives by its tributaries. It is the ocean then which must fill up the void; a portion of the

Fig. 24.—The Straits of Gibraltar.

current which flows from north to south along the coasts of Portugal and Spain, enters by the Straits of Gibraltar, and spreads far into the Mediterranean in superficial sheets. Nevertheless, if this inland sea did not also send a counter-current to the Atlantic, it would sooner or later be changed into an immense plain of salt. Incessantly losing fresh water by evaporation, and always receiving salt water from the ocean, its liquid mass would become in the end completely saturated, and the crystals of salt would line the marine bed in

* Bégy, *Annales des Ponts et Chaussées*, 1863. Vigan, ibid. 1866.

ever increasing layers. In order that the equilibrium of saltness between the two seas should not be interrupted, it is necessary that the Mediterranean should send its saltest waters to the Atlantic. This is, in fact, what takes place. Besides the lateral eddies that occur along the shores on each side of the current coming from the Atlantic, a Mediterranean counter-current flows below the lighter superficial waters, and takes its direction towards the ocean. This

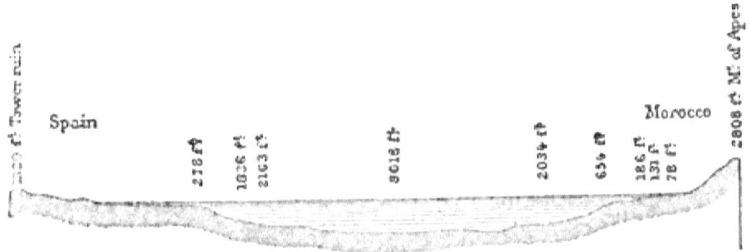

Fig. 25.—Profile of the Straits of Gibraltar.

submarine river, which passes the Straits of Gibraltar to be lost in the open sea, is, as chemical analyses have shown, a current of heavy water, almost saturated with salt. Thus, an exchange is accomplished through that narrow passage; the Atlantic gives to the Mediterranean the waters which it needs, and receives in return its superfluity of salt to diffuse through the ocean. The sea endeavours incessantly to re-establish its constantly disturbed equality at the boundary of the two marine basins, at a depth of about 546 fathoms.

This harmony of the forces of nature is shown in a still more striking manner at the entrance to the Red Sea. This elongated gulf, which is nearly 1480 miles in length from the Strait of Babel-Mandeb to Suez, receives from the atmosphere and the bordering countries so slight a quantity of water that it may be considered as absolutely nothing. It rains but very rarely over the sheet of water lying between the two deserts of Egypt and Arabia, and not a single torrent brings down its waters to it. The Red Sea is therefore only an immense basin of evaporation, and the annual loss is all the greater that the rays of the sun shine almost always from a cloudless sky. The portion of fluid transformed into vapour is estimated at about eight-tenths of an inch per 24 hours; that is to say, nearly 23 feet per year, so that if the gulf was completely closed, the water, whose mean depth does not exceed 220 fathoms, would be entirely dried up in the space of 60 years. Owing to their higher level, the waves of the Indian Ocean are carried into the Arabian Gulf by the Straits of Babel-Mandeb; and this flow, superficial or submarine, must make itself felt with all the more force, because during eight months of the year the winds blow from the north to the south, precisely in the axis of the Red Sea, and would thus tend to empty the gulf, if the laws of gravity permitted. But whatever be the swiftness of the current coming from the Indian Ocean, a portion of its water evaporates on the way, and, in consequence, the liquid mass, diminished by a certain quantity from evaporation, must become salter and salter in proportion as it advances to the north. In fact, it has been established by direct analyses, that the quantity of salt contained in the same volume of water increases gradually from Aden to Suez. From a little more than 39 parts in a thousand at the entrance to the gulf, it rises to 41 and even 43 parts in the thousand at the northern extremity.* Dr Buist, a scholar of Bombay, has calculated that if the Red Sea did not return to the ocean the salt that is concentrated there in consequence of evaporation, it would end in being changed into a solid mass of salt in a space of time certainly less than three thousand years, and perhaps in only fifteen or twenty centuries.† Now the Red Sea has already existed for thousands and thousands of years, and its waters (more salt than those of other seas, it is true) are still very far from being in a state of saturation. We therefore come to this inevitable conclusion, that a very salt submarine current flows through the Straits of Babel-Mandeb into the Indian Ocean, in an opposite direction, and below the superficial current which supplies the Arabian Gulf. As in houses, each door serves at the same time as a passage for two contrary currents, that of the warmer and lighter air which escapes above, and that of the colder and heavier air penetrating below, so in the seas, each strait is traversed by two streams, different in temperature and in their saline contents.

All these phenomena of exchange, which occur in such a striking manner at the entrance to the Red Sea, the Mediterranean, and the Baltic, are reproduced in the vast space of the seas wherever the equilibrium of level, warmth, or saltness, is disturbed by any cause whatever. Thus the Atlantic, much better supplied than the South Sea as regards rains and affluents, is nevertheless not more elevated; and on its side the Pacific does not contain a greater quantity of salt than the other oceans. On all parts of the planet, seas bathing the

---

* See above, p. 23. † Maury, *Geography of the Sea.*

shores of countries most diverse in appearance and geological formation, have a tendency to resemble each other in their composition, saltness, and in most of the other phenomena of their waters. The currents are the great agents in producing this equilibrium in the seas; but by their very mobility, their dependence on the seasons, winds, configuration of the coasts, and finally, by reason of the submarine part of their course, they are exceedingly difficult to observe in a systematic manner,—and among the numerous general and partial currents, there is not a single one, not even the Gulf-stream, whose normal course can be traced with complete precision. Happily, scientific observations are now being multiplied over all the seas; they add to and unite with one another; and, little by little, approach the truth by approximations which result from the comparison of facts. Every new sounding, every new thermometrical reading, is an acquisition to science, and allows us to follow with a clearer eye the complicated circulation of the waters in the immense labyrinth of the ocean.

BOOK III.—THE TIDES.

CHAPTER XI.

OSCILLATIONS OF THE LEVEL OF THE SEAS.—THEORY OF THE TIDES.

ANOTHER movement which keeps the waters of the sea in a constant agitation is that of the tides. While the currents carry the waves from one pole to the other, and stir the very mass of the ocean, the tides incessantly modify the level by the alternations of ebb and flow, which they impart to its waters. They raise or depress without relaxation the mass of waves on all the shores of the globe; the strand, which by turns they invade and lay bare becomes debatable ground between the two elements, and successively forms a part of the oceanic basin, and the continental surface. Twice a day vast plains of sand like those of Mount St Michael are invaded by the waves, deep bays are formed far into the land, and barks glide with sails spread above the path which the pedestrian has just quitted. Twice a day the same tidal wave causes the waters brought to it from the continents to return back again, transforms simple rivulets into large rivers, changes basins filled with mud into vast inland harbours, and carries fleets of ships over sandbanks and hidden rocks. Six hours afterwards all is changed. The tidal ports are strewn with ships stranded and lying in the mud, the mouths of rivers allow their islands of alluvium to emerge, and great bays are no more than plains of sand. Thus the outline of continents incessantly changes in appearance; the girdle of estuaries and ports, beaches, rocks, and sandbanks which surround their coasts, continually alter, and change the geography of the shores in the same proportion. Besides, movements so considerable cannot occur without being accompanied by very powerful currents, flowing alternately from the open sea towards the coast, and from the coast to the open sea, and contributing greatly to the general circulation and mingling of the waters in the ocean. The influence which the ebb and flow of the tides exercise indirectly on the commerce and civilization of nations is immense; it is to these movements of the sea that England owes in great part her power and glory.

In all times the people dwelling on the borders of the ocean have understood, without being able to account for it, that the alternate phenomena of ebb and flow depend on the position of the moon and sun relatively to the earth. The coincidences that they saw renewed each day between the movements of the tides, and those of the large heavenly bodies, could not leave them in any doubt of this. Sailors and fishermen, accustomed to look to the sky for the signs of the weather, and indications of the route which they ought to follow, had no trouble in ascertaining that the return of every second tide corresponds exactly to the passage of the moon over the same degree of the heavens; that is to say, to the commencement of a new lunar day. Following the phases of the moon, at new, half-moon, or full, they saw the tides change in a regular manner, and become successively higher and higher, and afterwards, from day to day, lower, till the end of the lunar month. Finally, the movements of the sun also announced to them beforehand the approaching state of the waves, for the equinoxes of March and September are always accompanied by very high tides. These coincidences between the phenomena of the sea and the movements of the moon and sun are so striking, that all barbarous maritime tribes have remarked them, and have rudely symbolized the idea in their songs. Thus the Scandinavian *sagas* represent Thor, the god of winds, blowing the water with a horn which he plunges into the depths of the ocean, and by his powerful breath causing the waves to rise and fall by turns. What can this strange legend signify, if not that the regular oscillations of the tide depend on the cosmical forces to which the planet itself is subject?

Nevertheless, these symbolic tales of the ancient Scandinavians are far removed from that scientific theory of the tides, which the researches and sagacity of Newton and Laplace have established. Even Pliny, when he affirmed clearly that the tides are due "to the combined influences of the sun and moon," restricted himself to summing up in precise terms what all the dwellers on the shores of the ocean knew; but he could not explain in what manner this influence was exercised. The explanation of the mysterious phenomena of the periodical swelling of the waters could only be attempted in modern times, with the aid of the knowledge obtained by astronomers on the motion of the celestial bodies, and with the powerful means of investigation, which mathematicians have supplied them with. Kepler first indicated the course to be followed; and Descartes, and then Newton, each gave his theory explaining the tides, the one by pressure, the other by the attraction exercised by the sun and moon on the mobile waters of the sea. It is the latter theory, that of Newton, which was developed later, much modified by Bernouilli, Euler, and Laplace, and which Lubbock, Whewell, Chazallon, and so many other natural philosophers have since compared with observations made on the shores of the ocean. Being very satisfactory in most respects, it is now very generally accepted; but it still has eminent opponents, among whom F. Boucheporn* must be named; many of the secondary facts are still to be elucidated, and many local phenomena are not yet understood. To follow the tides in their progress and fluctuations across the seas, it is not sufficient to know the laws of gravitation, and to calculate with the most rigorous exactitude the movement and position of the heavenly bodies; one must also know all the facts relative to the movements of fluids, and know how to apply to all their phenomena of acceleration, retardation, increase, interference, and equilibrium the most complicated and most minute formulæ of high mathematics. It would also be indispensable to know every fact respecting the form of the shore, and the inequalities of the bed of the sea.

Reduced to its principal elements, the theory of tides set forth by Laplace, and generally adopted since, is very simple. The earth is not an isolated body in space; it is attracted by all the nearer heavenly bodies, and it is indeed in great part this force of gravitation which causes it to turn round the sun, and retains the moon as its satellite. Let us imagine for an instant the earth to be covered with water over all its surface, and subject to the attraction of the moon alone. This superficial part of the planet would be more strongly attracted than the solid portion, since it is nearer to the moon which attracts it; and owing to the facility with which liquid particles glide one over the other, it would swell, so to say, towards the moon till its weight would be in equilibrium with the attracting force. It would then form an intumescence, the summit of which would be exactly on the ideal line which unites the centre of the earth to that of the moon. On

* *Philosophie Naturelle*, pp. 1—205.

THEORY OF THE TIDES.

the other side of the planet, according to the general theory, the waters ought to swell in a corresponding wave, and that from a precisely contrary cause. The liquid strata on this part of the earth being further from the moon than the solid kernel, are less attracted than it, and in consequence must remain slightly behind, thus forming a new intumescence, the summit of which will be found on a prolongation of the line uniting the planet with its satellite. Considered

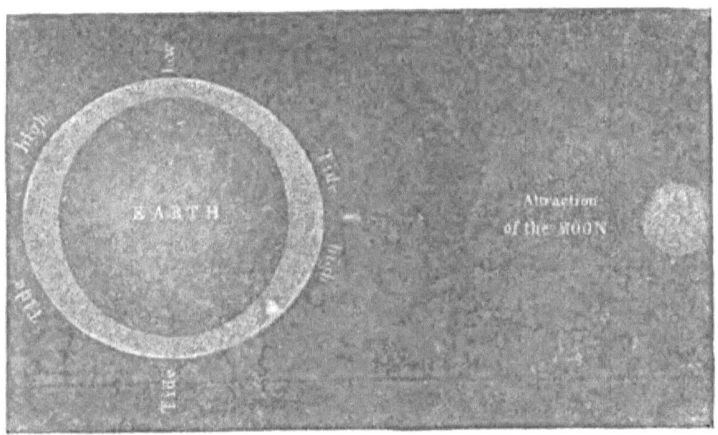

* Fig. 26.—Lunar tide.

as a whole, the mass of marine waters would thus assume the form of an ellipsoid, having its greater axis directed towards the moon, which is the centre of attraction. It results from this, that the tide ought to be nothing at all, or very slight, at the poles; since in its revolution, the moon while moving to the north and south of the equator maintains itself at the zenith of tropical or sub-tropical regions.

If the earth remained immovable these two waves would advance slowly, following the course of the moon; but in consequence of the rotation of the earth, they ought to move rapidly in pursuit of one another over its circumference: the wave of the greatest attraction moving incessantly over the part lighted by the rays of the moon, while the wave of the weakest attraction is propagated from the other side of the earth on the part furthest from the satellite. In

* This illustration, as well as figs. 27 and 29, have been borrowed from the fine work by M. Amédée Guillemin, entitled *Le Ciel*.

the space of a lunar day, that is to say, within the 24 hours 50 minutes during which the earth has successively presented all parts of its surface to the planet which accompanies it, the two waves ought each to accomplish a complete circuit around the globe, and

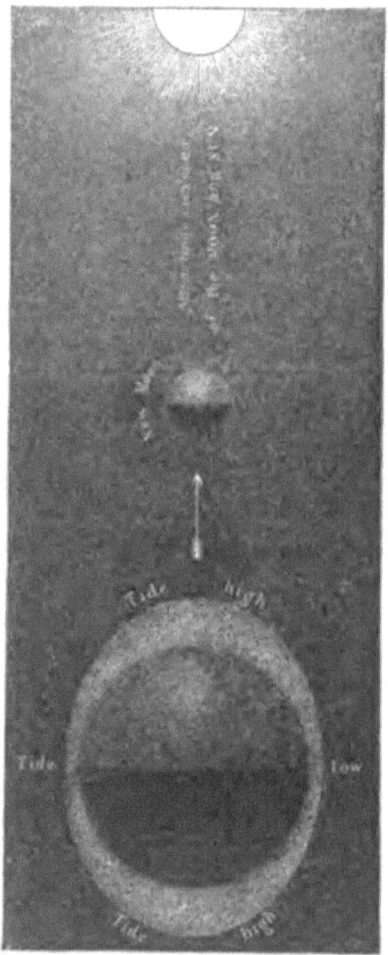

Fig. 27.—Syzygy tide, during the new Moon.

each should have a total duration of 12 hours 25 minutes. This is, in fact, what takes place over all seas. As to the numerous variations presented by this phenomenon, in its height and the precise

THEORY OF THE TIDES.

moment of its appearance, they depend on the obstacles of every kind which the rocks, islands, continents, oceanic currents, and winds oppose to the free circulation of the waters.

Nevertheless, the moon is not the only heavenly body whose attraction is manifested in a sensible manner on the waves of the ocean. The sun, which draws the moon in its immense orbit across the heavens, is near enough to our planet to raise the liquid particles of our

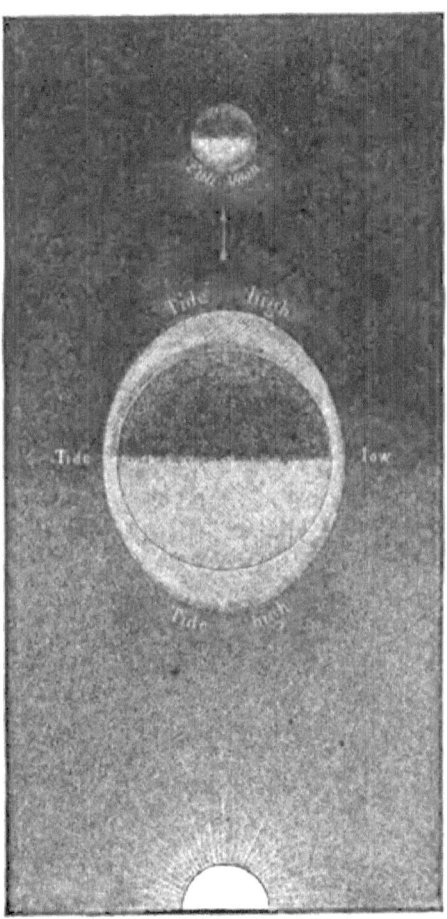

Fig. 28.—Syzygy tide, during full Moon.

ocean also. The total attraction exercised by the sun on the earth is even 162 times greater than the total attraction of the moon, and

98   THE OCEAN.

in consequence it would raise the tides into real mountains as high as the Cevennes,* if the true cause of the tides was not to be found in the difference of attraction exercised on the waters of the different parts of the earth. The distance from the moon being equal to 60 terrestrial radii only, the action of the satellite is much stronger over the nearer oceanic regions than over the waters situated thousands of miles further off. The sun, on the contrary, acts nearly in the same manner on the watery particles of the whole surface of all the seas. According to the results obtained by the calculations of mathematicians, the attractive force exercised by the sun in elevating the waves is, as compared to that of the moon, in the proportion of about a third.

Two tidal waves, the lunar wave and the solar wave, are thus raised on the surface of the sea. They ought to revolve, the one in

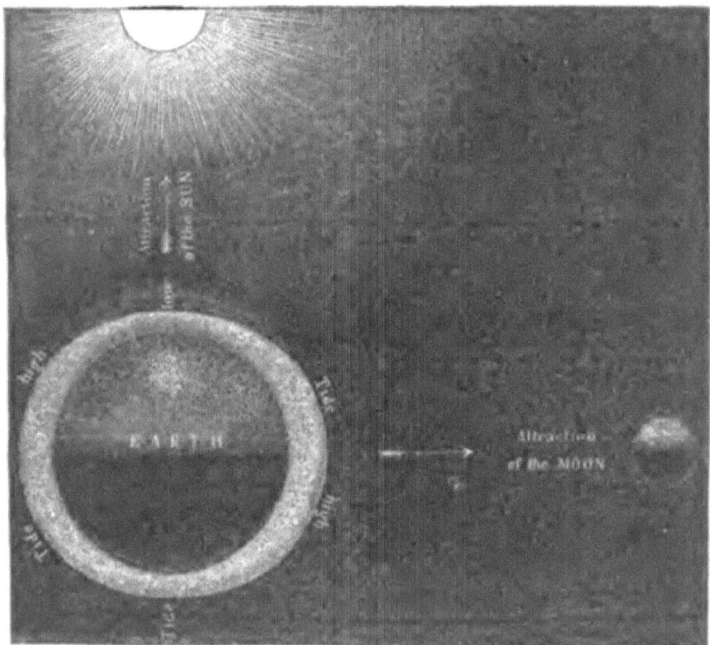

Fig. 29.—Tide during Quadrature.

the space of 24 hours 50 minutes, and the other in 24 hours. But these two waves so distinct in their origin are not separated in their

* 5000 to 6000 feet high.

course around the globe; owing to the incessant mobility of the waters, they mix and are confused, and it is by calculation alone that we can discriminate in their common mass the part that is to be referred to each of the two heavenly bodies. These two united intumescences move together around the earth in a direction from east to west; that is to say, in the opposite direction to the rotation of the globe. Serving thus as a drag upon the planet, they must in the long run lead to that slackening of its speed, which the calculations and deductions of Meyer, Tyndall, Joule, Adams, and Delaunay lead us to consider as inevitable.*

When the moon, called new, turns its dark face towards us, and is thus in nearly the same direction as the sun relatively to the earth, the attractions of the two great celestial bodies join together, and the two tidal waves, raised at the same time towards the same point of space, are exactly superposed. They form those tides of syzygy or high water, called spring-tides, which rise to such great heights along our shores. At the time of full moon, that is to say, when the satellite, entirely lighted, is in direct opposition to the sun, new tides of syzygy not less elevated than the first are formed; for under the influence of the heavenly bodies situated opposite to each other a double intumescence is simultaneously produced on both sides of the earth. During none of the other phases of the moon does this coincidence exist; at the time of quadrature, the two great movements of the waves oppose one another, and the tidal wave, which represents then the lunar wave diminished by the entire solar wave, is less elevated than during the other phases of the moon. If the two attracting forces were equal in power, the neutralization of the tide would be complete, and the level of the sea would remain undisturbed.

To give an idea of the fluctuations which occur during the course of an entire tide, under the influence of the heavenly bodies, and which are variously modified by the atmospheric currents, the form of the coast, and inequalities of the bed of the sea, we borrow the following figure from Beardmore.

The periods of the tides are exactly those of the bodies which raise them. The semi-diurnal period of 12 hours 25 minutes is comprised between the passage of the moon over the two opposite meridians of the earth. The diurnal period, during which the ocean swells and subsides twice, corresponds exactly to the duration of one apparent rotation of the satellite around our planet. There is the

* See in Vol. I. the section entitled, *The Earth in Space.*

same coincidence for the semi-monthly period; the return of the spring-tides occurs from fortnight to fortnight with the return of the full or new moon, and the monthly period is completed when the series of lunar phases recommences. Nor is this all; the tides have

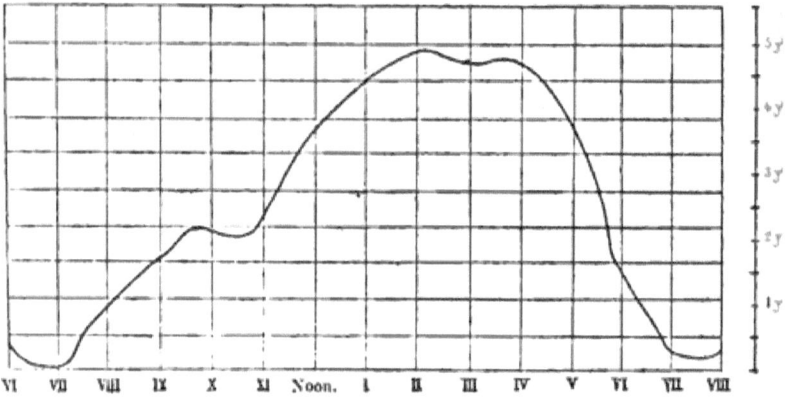

Fig. 30.—Tide at Southampton, 2nd August, 1859.

also their semi-annual period, from the equinox of March to that of September, for the sun being then directly above the terrestrial equator, exercises a stronger attraction on the liquid masses, and the waves of the spring-tides rise to a greater height than usual. Finally, an annual period is marked for the tides by the epoch when the earth is nearest the sun. This epoch falls during the winter of the northern hemisphere, and it is then indeed that the spring-tides rise with most force on the coasts of our continents.

Thus the phenomena of the tides are intimately connected with the celestial movements, and every change in the relative position of the bodies which attract our planet, manifests itself by a corresponding change in the level of the seas. Knowing beforehand the route which the earth follows in space, astronomers foresee thereby even the future oscillations of the wave, and can trace their curve for centuries to come. Nevertheless, it must be admitted, this curve is only true in theory; for if the tides in their origin be due to astronomical causes, they are also subject to variations from terrestrial phenomena. Like the winds, currents, and all the other manifestations of planetary life, they present incessant variations, and are, so to say, in a continual genesis.

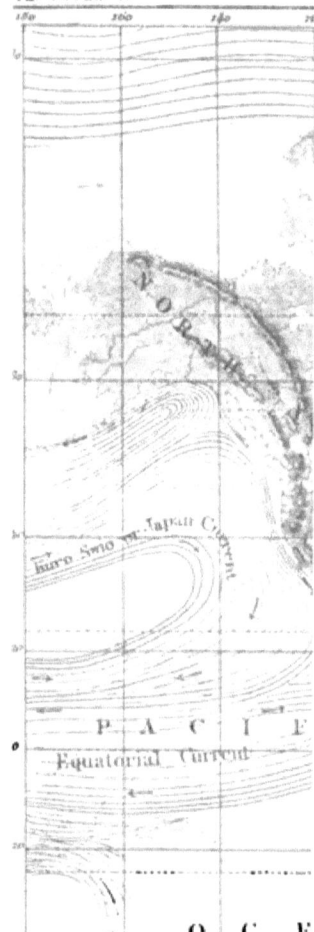

CHAPTER XII.

THEORY OF WHEWELL ON THE ORIGIN AND PROPAGATION OF TIDAL WAVES. —ORIGIN OF THE TIDE IN EACH OCEANIC BASIN.—"ESTABLISHMENT" OF PORTS. —"COTIDAL" LINES.

THE English natural philosopher, Whewell, who during long years made laborious researches on the phenomena of ebb and flow, was the first to apply the name of "cradle of the tides" to the great continuous sheet of water which covers almost all the surface of the southern hemisphere. It is in this vast basin, of which all the other oceans are mere ramifications, that the combined attraction of the sun and moon would first raise that wave, which from shore to shore dashes at length against the coasts of Greenland and Scandinavia. It is there that the water, a few instants after the passage of the moon over the meridian, would itself attain the level of its highest elevation, and would form that first regulating intumescence, which the surface of all the seas would obey one after the other, as a cord shaken at one of its extremities oscillates to the other end in rhythmical vibrations.

According to this theory, the tidal wave circulates incessantly throughout the Antarctic Ocean, to the south of the extremities of the three continents of Australia, Africa, and South America. It follows from east to west the apparent course of the moon, and thus describes a real orbit round the earth similar to that of the celestial bodies. Even in the central Pacific and the Indian Ocean, the tide obeys this normal impulse towards the west. It strikes the coasts of Australia and New Guinea almost simultaneously; then thirteen or fourteen hours afterwards it dashes on the eastern coast of Africa, from the bank of Lagullas to Cape Guardafui; finally, seven or eight hours later, the coast of South America is struck in its turn from Terra del Fuego to the estuary of La Plata.

To the north of those large oceanic tracts of the South Sea, the tides, not having the same facilities for developing themselves in a normal manner, would be obliged to change their direction. But in spite of this deviation, they would not the less be, Whewell thinks, continuations of the primitive swelling. Arrested by the American continent, which bars its passage, the tidal wave would rebound towards the north, and follow the contours of the oceanic valley, like a torrent enclosed in a mountain gorge. Striking the coasts of America and those of the Old World, under the same latitude, at the same time, and at an equally oblique angle, it reaches almost simultaneously, on either side of the Atlantic, the Bay of Fundy and the Irish Channel, where its highest known elevation is observed. The tidal wave accomplishes this passage of about 6000 miles, from the Cape of Good Hope to the British Isles, in about fifteen hours. But its entire voyage, from the centre of the Antarctic Ocean, must have lasted more than a day, and in consequence of the gradual slackening of speed of the waters on the shores of Great Britain, it is only after two days and a half that the tidal wave reaches the mouth of the Thames. Thus the moon would have had time to raise five successive tides in the Pacific Ocean before the motion of the liquid mass would have been propagated to the entrance of the North Sea.

Such is the theory which the labours of Whewell have caused to be long considered as the very expression of truth. Nevertheless, it is not certain that things occur in this way. In fact, it is ascertained that in each oceanic basin the tide seems to start from the centre, and to be propagated in all directions parallel to the general direction of the coasts. We may naturally conclude from this, that each great division of the ocean, considered as an isolated sea, is really the cradle of the tides which break upon the surrounding shores. What confirms this idea, too, which appears so probable at first, is that the various oceans are separated from one another by spaces where the regular tide is hardly perceptible. Thus between the South and North Atlantic, whose precise boundary may be defined by the promontory of St. Roque and Cape Verde, there exists a wide zone where the tide hardly changes the maritime level more than about 23 to 27 inches, as at the islands of Ascension and St. Helena. Besides, according to the theory of Whewell, the tidal wave on the coasts of the Argentine Republic and Brazil, ought to propagate itself from south to north; whilst, on the contrary, the movement proceeds from north to south, from Pernambuco to the mouth of the La Plata.* When we see a tidal wave rise off the bank of Newfoundland, in the deepest part of the Northern Atlantic, it is not therefore necessary to consider this as the same wave which twelve hours before was raised near the bank of Lagullas at the en-

* Fitzroy, *Adventure and Beagle.* Appendix to Vol. II.

trance to the South Atlantic. It is perhaps better to regard the
oscillations which occur at the same time in both hemispheres, as
coincident but independent phenomena.

Nevertheless, in each isolated basin the movements of the sea are
much as Whewell has described them. On the coasts of France
and the British Isles the tide certainly comes from the open sea, and
in its progress along the shores, the original motion which the attraction of the sun and moon produced in the middle of the open sea continually decreases. On penetrating into the shallower seas which
surround Ireland and Great Britain, the tidal wave gradually
slackens. After having struck Cape Clear and the promontory
of Land's End, it is propagated with such slowness around the two
islands, that 19 hours elapse before it arrives at the Straits of
Dover, where it meets with another wave newer by 12 hours,
which has come by the shorter route of the Channel. Whence
comes this slackening of the wave? The researches of astronomers
and natural philosophers inform us, that the speed of the tidal wave
is proportioned to the depths of the ocean; driven by an equal
force, the circumference of a wheel turns the faster the greater its
diameter; in the same way the tide hastens or slackens its movement, according to the depth of the watery mass which it traverses.
In those latitudes where the bed of the ocean is 5000 fathoms
from the surface, the speed of the wave is about 528 miles an hour;
where the depth is only about 50 fathoms, the tide is not propagated
more than about 60 miles in the same space of time; finally, when
the bottom is at about 5 fathoms below the marine surface, the movement of the waters is greatly retarded, and does not exceed 15 miles
per hour, that is to say, 440 yards per minute.

In consequence of the delay which the tidal wave experiences, the
"establishment," that is to say, the time which elapses between the
passage of the moon over the meridian and the moment of full
tide, varies singularly in different ports situated near each other.
Thus while at Gibraltar there is usually a coincidence between the
astronomical and marine phenomena, and the establishment is reduced in consequence to zero; this interval is about an hour and
15 minutes in the port of Cadiz, and 4 hours at Lisbon. At
Bayonne, as at Lorient, it is 3 hours 30 minutes, at the mouth
of the Gironde and at Cherbourg it is 7 hours 40 minutes, at
Havre 9 hours 15 minutes, at Dieppe 10 hours 40 minutes, at
Dunkirk 11 hours 45 minutes. The establishment varies on every

shore according to the speed of propagation of the tide across the open seas and in the gulfs and estuaries.

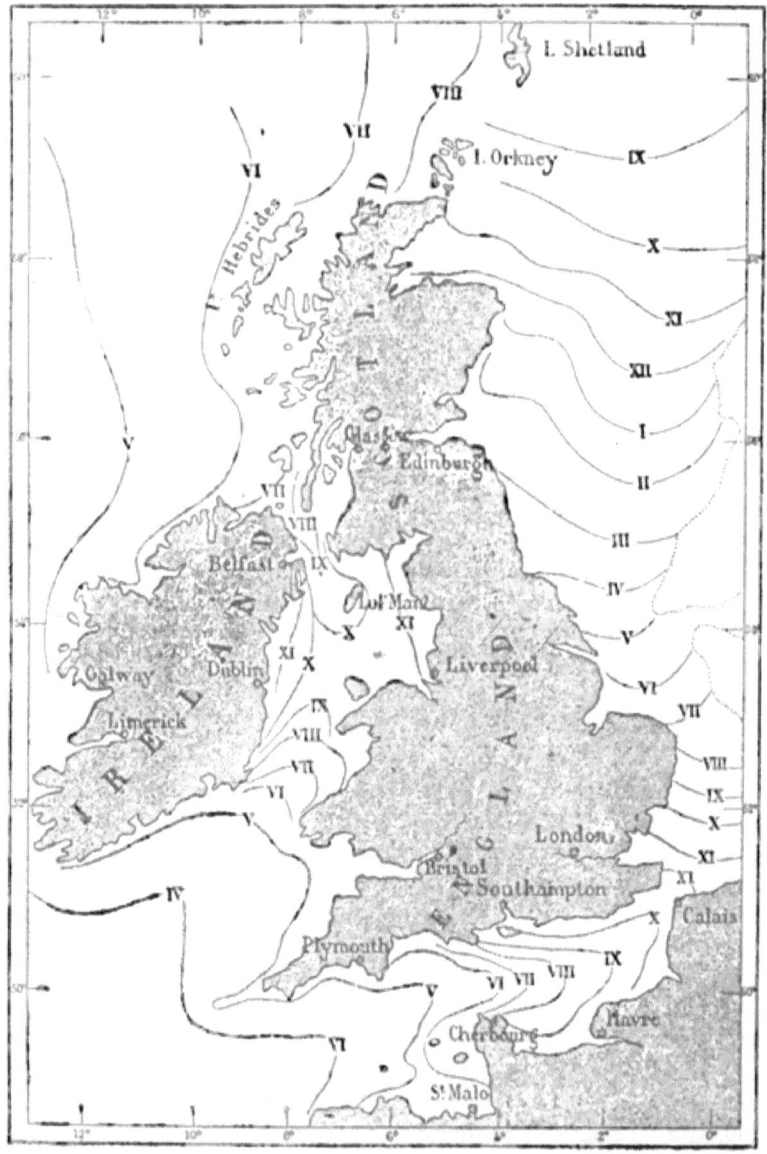

Fig. 31.—Cotidal lines of the British Isles.

The sinuous line which unites all the points in the ocean where the full tide occurs exactly at the same hour, has received from Whewell the name of *cotidal* line; it indicates the curve which the crest of the tidal wave forms at any one moment on the surface of the sea. It is around the British Isles that these lines of simultaneous swelling or of equal establishment, have been most carefully traced. By calculation and direct observation, that part of the oscillation on the mobile and almost always agitated surface of the sea, which is to be referred to the phenomena of ebb and flow, has been detected; and much more exact maps of these swellings and depressions, which are invisible on the open sea, have been drawn, than of the vast continental regions which are at present but little known. Thanks to the labours of Whewell, Airy, Lubbock, and Beechy, one can now follow the whole series of cotidal lines which succeed one another from hour to hour around these two great islands, from the crest coming in from the open sea, at the entrance of the English Channel and the Irish Sea, 4 hours after the passage of the moon over the meridian, to the swelling, which 19 hours later reaches to the south of the German Ocean and penetrates into the funnel of the Straits of Dover, where it meets the other tidal wave coming directly by the Channel. The general form of these curves demonstrates in a striking manner that the speed of propagation of the tide is in proportion to the depth of the seas. Everywhere we see the cotidal lines develope their convex part above the deeper valleys of the marine bed; everywhere we see the wave slacken its speed in the neighbourhood of shallow rocks and shores. One could even by an inspection of these lines of equal intumescence indicate exactly those parts where the lead would descend lowest; so intimate is the connection of cause and effect between the depth of the sea and the progress of the tide.

CHAPTER XIII.

APPARENT IRREGULARITIES OF THE TIDES.—EXTRAORDINARY SIZE OF THE TIDAL WAVE IN CERTAIN BAYS.—INTERFERENCE OF EBB AND FLOW.—DIURNAL TIDES. —INEQUALITIES OF SUCCESSIVE TIDES.

INNUMERABLE are the apparent irregularities which occur in the phenomena of the tides, in consequence of the inequalities of the submarine surface, the thousand indentations of the shore, and the alternations of winds and currents. Though the cause of the movement be the same everywhere, we can still say that at no point of the

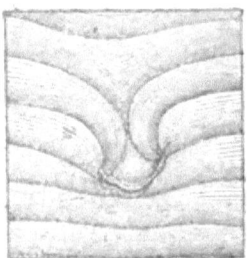 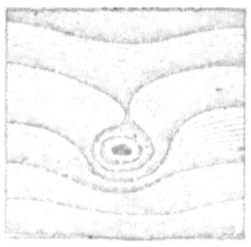 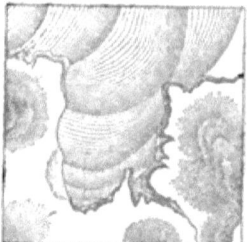

Figs. 32—34.—Irregularities in the curves of the Tidal waves resulting from the form of the sea-bed, projecting rocks, &c. (after Lubbock).

sea do the ebb and flow present a perfect agreement in their progress. Each promontory, each islet, each rock is bathed by waters having a distinct rule in the propagation of their tides; every obstacle which breaks the regular course of the oscillations modifies the whole of the graceful curves which bend around it. The above figures, borrowed from Lubbock, give an idea of these variations in the march of the waves.

The difference which most strikes the minds of navigators and inhabitants of the coast is that of the height of the tides. In one part of the coast the tide hardly makes itself felt, even during the equinoctial syzygies; while elsewhere every tide is a real deluge, spreading as far as the eye can see over vast tracts, which emerge again at the time of ebb. This astonishing contrast in the amplitude of the tides results from differences of speed in the progress of the oscillations in the seas and bays of the coast-line. In fact, the great swelling caused by the heavenly bodies may be considered as formed of a great number of successive waves occupying a considerable breadth on the surface of the sea. In the open ocean, all these waves move with great speed; but in proportion as they approach the shores, they slacken their movement, and consequently must gain in height what they lose in rapidity. From the mere sight of a tidal chart we can affirm that the tide will rise several feet high in all the gulfs where we see the cotidal lines crowded together, in consequence of the gradual retardation of the wave of intumescence.

In this respect, facts fully confirm theory. The Gulfs of Bengal and Oman, the Chinese Sea, the indentations of the eastern coast of Patagonia, the Bay of Panama, that of Fundy, between New Brunswick and Nova Scotia, the Channel and the Irish Sea, are parts where the waves of equal intumescence follow each other very closely, and it is there too that a greater extent of coast is alternately covered and revealed by the tide. In the port of Panama the tides rise nearly 23 feet, concealing and discovering by turns an immense strand in their diurnal movements, while at hardly 37 miles distant on the other coast of the isthmus the ebb and flow are scarcely perceptible.

In the Persian Gulf and the Chinese Sea the amplitude of the equinoctial tide is nearly 36 feet at the extremity of the gulfs. In the mouth of the Severn and the French bay of Mount St. Michael the difference of height between the spring-tides and low-water is from 45 to 48 feet. To the south of the American Continent, in the Gulfs of San-Jorge and Santa Cruz, at the entrance of the Straits of Magellan, Fitzroy has measured tides of from 48 to nearly 66 feet high; finally, in the Bay of Fundy, so well calculated, by the contour of its coast and the surface of its bed, to retard progressively the march of the tide, the difference between high and low water, which is about 9 feet at the entrance, gradually increases to nearly 69 feet towards the extremity of the channel. This is probably the part of the coast where the regular oscillations of the waters are accomplished in the grandest manner. Twice a day immense neutral shores, which are neither land nor sea, change into deep gulfs, and stranded ships rise and float with sails spread, whilst towns lost in the

* De Boucheporn, *Philosophie Naturelle*.

interior of the land find themselves seated on peninsulas invested by the sea. At St. John's, New Brunswick, a cascade is seen to glisten

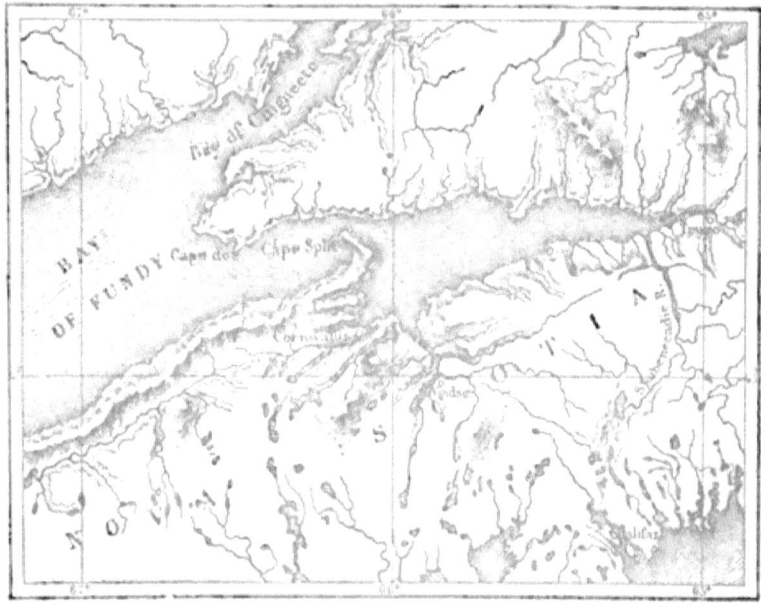

Fig. 35.—Bay of Fundy.

at the bottom of the port at low water; but when the tide reaches the foot of the cliff, the height of the fall gradually diminishes, and it is at last entirely drowned in the salt waters, which, spreading far

Fig. 36.—Mouth of the Avon: after Beardmore.

over the upper terrace, permit vessels to penetrate into the natural basin formed above the cascade.

PORTS REMARKABLE FOR HIGH TIDES. 109

Similar phenomena occur in the two bays of Mount St. Michael and the Severn. There, too, rivers and rivulets are periodically changed into gulfs; there, too, the harbours are tidal ports, where ships, with the exception of those which are enclosed within the basins, lie on their sides in sand or mud at the time of low water. In the same way the space extending between Noirmoutiers and the coast of

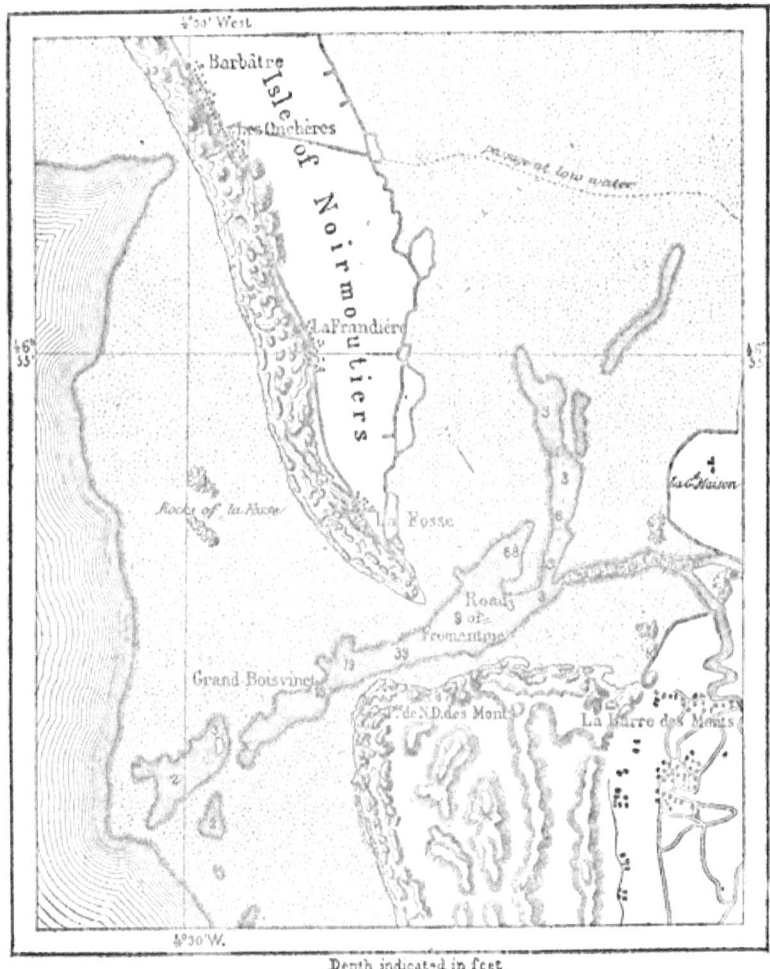

Fig. 37.—Straits of Noirmoutiers.

La Vendée is alternately an isthmus and a strait; a high road traversed by vehicles winds through the sandy plain between pools of water, and a few hours afterwards vessels with sails spread pass over the same route. Sailors are often seen walking quietly on the shore at a slight distance from their stranded vessel, or else digging in the ground in search of shells; but let the distant rolling of the tide be heard, and in the space of a few seconds the crew is on board, preparations are made for a new embarkation, and the vessel, raised by the tide, sails rapidly over the sea.

It is in the bay of St. Michael on the western coast of Europe that the rising tide presents the grandest spectacle, for in the centre of the bay rises a black granitic rock, "abbey, cloister, fortress, and prison" at the same time, which by its abrupt precipices and its "titanic pile, rock upon rock, century after century, but always dungeon over dungeon," contrasts with the dreary extent of the shore.[*] At low water, the immense sandy plain, above 150 square miles in extent, resembles a bed of ashes. But when the tide, swifter than a horse at full gallop, rises foaming over the scarcely perceptible slope, a few hours are sufficient to transform the whole bay into a sheet of greyish water, penetrating far up the mouths of the rivers as far as the quays of Avranches and Pontorson. At the ebb, the waters retire with the same speed to nearly $6\frac{1}{4}$ miles from the shore, and lay bare the great desert strand, which is intersected by the subterranean deltas of tributary rivulets, forming here and there treacherous abysses of soft mud, into which travellers are in danger of sinking. At the time of spring-tides the liquid mass which penetrates into the bay is estimated at more than 1470 millions of cubic yards, and even at neap-tides the deluge, which pours over the beach twice in the four-and-twenty hours, is not less than about 765 millions of cubic yards.[†] Is it astonishing that such torrents should have been able in former times, when driven by tempests, to break through the chain of sand-hills which protected the rocks of Tombelène and St. Michael on the north, and to transform into sterile wastes the beautiful country and vast forests which extended to the foot of the peninsula of Cotentin?[‡]

Beechey's observations of the tides of the Channel and the Irish Sea cause it to be regarded as certain that the enormous amplitude of the ebb and flow at the mouth of the Severn, and in the bays of Cancale and St. Malo, arise, not only from the gradual elevation of the bottom, but also from the superposition of two waves, which encounter each other. In fact, the crest of the tide which penetrates

---

[*] Michelet, *La Mer*, p. 18.
[†] Marchal, *Annales des Ponts et Chaussées*, 1854.
[‡] See in Vol. I. the section entitled, *The slow Oscillations of the Land.*

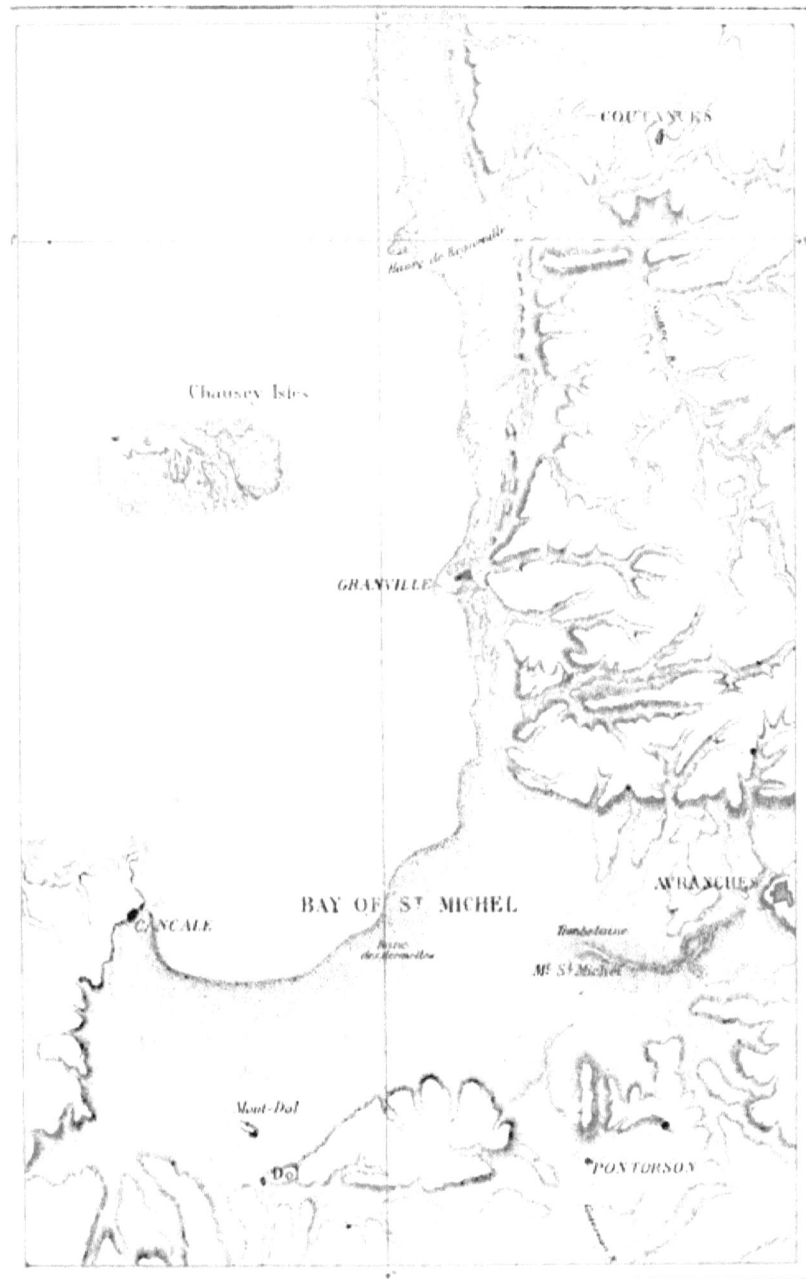

into the Irish Channel, meets at the end of the Gulf where the Severn discharges itself, another wave older by twelve hours, which has just made the entire circuit of Ireland. These two waves, united into one, take the common direction which results from their original impulsion, and flow together into the Gulf of the Severn. In

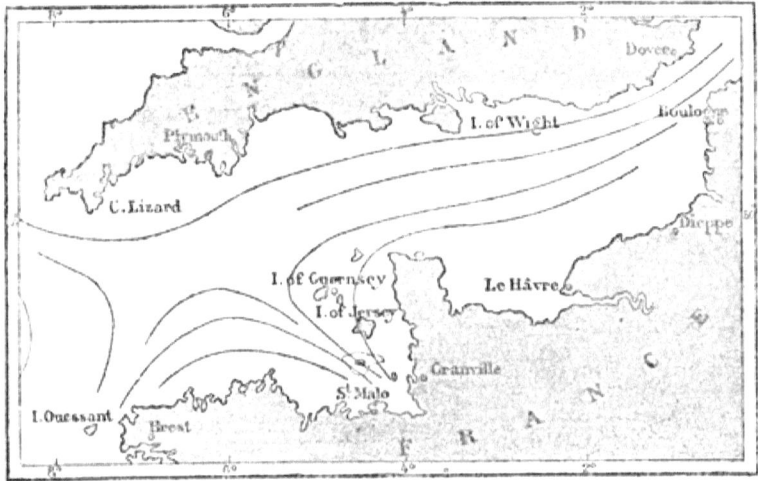

Fig. 39.—Tides of the English Channel.

the same manner, the tide which enters the Channel meets off Jersey with another wave, which has made the tour of the British Isles in twenty-four hours, and the two joining each other, dash their enormous liquid mass against the strand and rocks of Brittany.

If two tides coming from opposite points, and meeting at the time of high tide, are thus combined in one, they, on the contrary, neutralize and suppress each other, when the ebb of the one crosses the flow of the other. A phenomenon of interference occurs then comparable to that of two luminous vibrations extinguishing each other. Fitzroy was the first who pointed out a region of the ocean where contrary tides maintain the surface of the water in equilibrium. This region is the estuary of La Plata. At sight of this gulf, which is no less than 150 miles at the entrance, one would be tempted to believe, that the amplitude of the ebb and flow would be as enormous there as in the Bay of Fundy or the Gulf of St. Malo. But, on the contrary, the tides there are scarcely anything. The strong oscillations of the level that have been observed in that estuary, are due almost wholly to the regular breezes and the tempests, which depress the waves on one side and raise them on the other. Then, too, as the land winds generally predominate during the morning, and are replaced in the evening by the sea-breezes, the ebb and flow, obedient to the alternating impulses of the atmosphere, succeed each other every twelve hours; the tide rises in the afternoon and falls the next morning.* This apparent anomaly is easily explained by the meeting of high and low water at the entrance of the estuary. The tidal waves which flow to the south on the Brazilian side, and to the north on the side of Patagonia, do not strike the coasts at the same instant daily. They follow each other at an interval of several hours, and the lateral currents which diverge from them succeed one another at the mouth of the estuary of La Plata, so as to maintain the liquid mass at nearly the same level. At the moment when the ebb of the northern tide is about to occur, the southern flow takes place, the pressure of which, exercised in the contrary direction, prevents the waters from falling; then when a new tide from the coasts of Brazil presents itself, the surface of the sea is already lowered in the southern latitudes. The swellings would intersect each other, and on the line of interference the water would be subject to no oscillations.

It is probable that to causes of a similar kind we must attribute the formation of those diurnal, and always very slight, tides which occur at the mouth of the Mississippi, on the coasts of New Ireland, at Port Dalrymple in Tasmania, to the south of Australia, near King George's Gulf, in the Gulf of Tonquin, in the Bay of Bahr-el-Benat, in the Persian Gulf, in the White Sea, and in many other parts of the ocean. These slow changes of level, the ebb and flow of which each lasts twelve hours, present, like ordinary tides, the greatest diversity in their phenomena, according to the direction of the winds and the currents, the respective positions of the sun and moon, and the parts of the sea where this equilibrium of the waters is established. On the moving surface of the ocean, all the undulations, whatever may be their cause, are mixed and confounded, and in this ceaseless changing and mingling of the waves it is impossible to discern, without long and patient research, the part taken by each agent in disturbing the perfect repose of the sea-level. The problem can be solved in a general manner only, without taking account of details that have been as yet imperfectly observed. Thus, it is known, that in the port of Vera Cruz, and on the neighbouring

* Martin de Moussy, *Confédération Argentine*, t. i. p. 78.

coast, the winds have a marked preponderance, for they sometimes maintain the surface of the sea at the same level during whole days. At the mouths of the Mississippi, where the daily tide has a rise of little more than 14 inches, it is not less regular in its progress, and its total height each day represents exactly the difference of level between the two composing waves, which have crossed each other. Finally, the tide at Tahiti, nearly 12 inches high, is the result of many more oscillations; for four tides, coming from the four cardinal points, meet each other there, all differing in their speed and their hour of high water. It is not surprising that in the middle of this general intersection of the tides of the Pacific Ocean, that of Tahiti is almost completely neutralized.*

The Irish Channel, so well studied by Beechey, presents a very curious example of a perfect equilibrium of waters, and that almost opposite the Bristol Channel, where the sea rises and falls alternately above 48 feet. That part of the Channel whose surface remains at rest borders on the Irish coast not far from the little town of Courtown, to the south of Arklow. There, neither rise nor fall in the waters has ever been observed, though the currents of the ebb and flow run along the coast alternately, with a speed of nearly 4½ miles per hour. The point where the waters are always in equilibrium may be considered as a kind of "hinge" on which the tides turn. Their amplitude is greater and greater in proportion as they are distant from this tranquil region, to the northeast towards Holyhead and Liverpool, to the south-east towards Milford Haven and Bristol. In the North Sea, the meeting of high and low water, not far from the Straits of Dover, is marked by another centre of equilibrium, which seems to oscillate between the coasts of Holland and those of England, according to the atmospheric and marine currents, and the movements of the heavenly bodies. In this place, Hewitt has ascertained that the tide rises two feet only; and it is in this region where the waters keep almost always at the same level, that the largest and most numerous sand-banks are deposited.

It appears that the two tidal currents which meet near the Straits of Dover, the one coming directly from the Atlantic, the other from the North Sea, do not follow the centre of the Channel, and consequently do not encounter each other directly. The rotation of the earth which in the northern hemisphere displaces all moving bodies towards the right, causes each of the tidal waves to diverge in this direction. In the Channel the tidal wave, which is directly

* Fitzroy, *Adventure and Beagle*, Appendix to Vol. II. p. 290.

114 THE OCEAN.

propagated, constantly leans towards the right, that is to say, towards

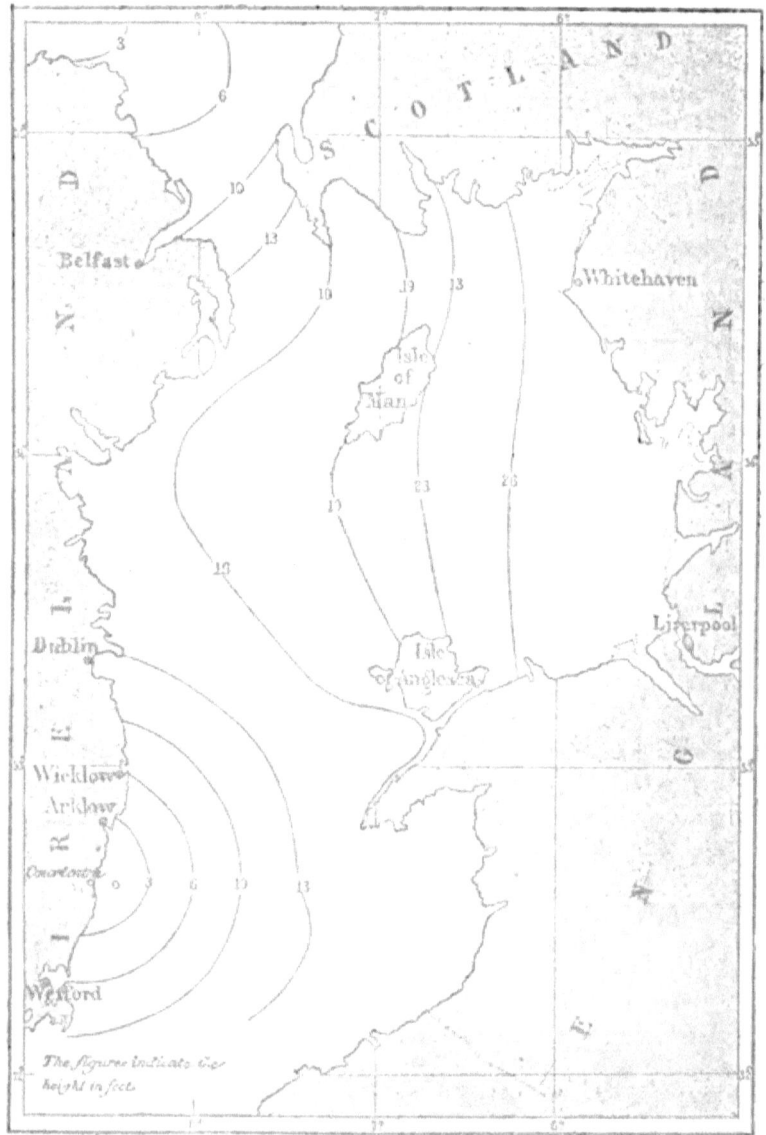

Fig. 39.—Height of the Tides in St. George's Channel.

the south; its force is therefore much greater on the coasts of France than on those of England, and when it has passed the Straits, it keeps its preponderance on the coasts of the continent as far as the mouths of the Meuse; the tide coming from the north, on the other hand, deviates likewise to the right and flows along the coasts of England. The crossing of these two contrary currents gives rise to numerous

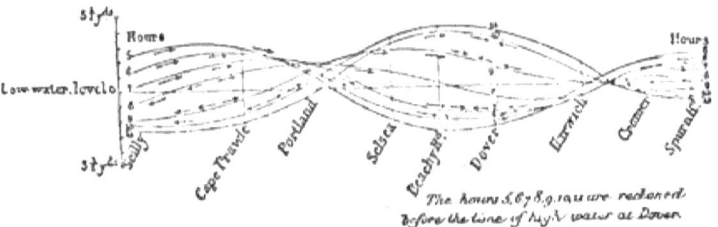

Fig. 40.—Crossing of the swellings of the tides in the English Channel and the North Sea, from the Scilly Isles to the mouth of the Humber.

gyratory movements off the coasts of France and Great Britain, the incessantly changing curves of which form a veritable labyrinth.*

In the roadstead of Havre the meeting of the tides results in a remarkable phenomenon, which is at the same time one of the most useful for navigation. Instead of falling immediately after having attained its point of highest tide, the sea remains steady for three hours, and thus permits vessels to sail all over the road, and to penetrate with ease into the port, floating constantly over deep water. The seamen saw in this fact a sort of miracle, before its true cause had been revealed. When the tide from the Atlantic rolls towards the east to the middle of the Channel, it is arrested in its course by the peninsula of Cotentin, and can only advance freely to the north of the Gulf, towards the mouth of the Seine. The marine level is thus more elevated at the centre than on its shores, and its waters are spread laterally towards the road of Havre and the other parts of the coast. At the time of low water, when the ebb prevails in the centre of the Channel, the inclination is changed; but before the waters of Havre can descend towards the central course of the Channel, which carries such an enormous mass of fluid to the ocean, they are kept back by the wave which, after having struck the Cape of Antifer, flows along the shores from north-east to south-west to the Cape of La Hève. Then, when the force of this partial tide fails, another river-tide, which has followed the coast of Normandy from St. Vaast to Trouville, still maintains the level, for a time.†

In almost all river ports, as we can easily understand, the ebb lasts longer than the flow, for the fluvial current neutralizes the tide during

* *Annales des Ponts et Chaussées*, 1863, first week. † Baude, *Revue des Deux Mondes*.

ing a shorter or longer period, and then adding to the ebb cannot but augment its duration.* A fact more difficult to explain is that, whilst in the greater number of ports remote from any river's mouth, the rising tide is shorter than the falling, numerous instances of the opposite are to be seen; and especially the port of Holyhead. According to the hypothesis generally adopted, this longer duration of ebb ought to be attributed to the rotation of the earth in the direction of west to east. The tidal wave being propagated in the contrary direction, that is to say, from east to west, would meet a certain resistance in the waters which are spread before it. It would rise up and become steeper and more rapid towards the west; while its other slope, that of the ebb, would lengthen itself towards the east. This will explain why the phase of the flow does not last so long as that of the ebb.

The inequalities, which are observed in certain parts between two successive tides, are likewise a strange and, in some respects, unexplained phenomenon. These various inequalities, now in the duration, and now in the respective heights of the two tides of morning and evening; or which even affect every oscillation in its entire course; arise in part from the declination of the moon, that is to say, from its varying distance to the south or north of the equinoctial line. But in many cases the differences between two successive tides are relatively enormous, and this explanation is not sufficient. Thus at Port Essington, on the northern coast of Australia, differences in height of nearly 4 feet between the oscillation of evening and morning have been observed. At Singapore, where the mean tide during the time of highest water is nearly 7 feet, the difference between two succeeding tides is sometimes nearly 5 feet. At Kurrachee the daily variation is no less, and in the Gulf of Cambay it attains to nearly 7 feet. At Bassadore, at the entrance of the Persian Gulf, the duration of one oscillation of the sea sometimes exceeds by two hours that which follows it; and, finally, it has happened at Petropaulowski, in the northern Pacific, that expected tides have never appeared at all. We can explain these singular anomalies only by the intersection of several reflex waves, diurnal and semidiurnal, which interfere with one another; and the confused oscillations of which are produced by the meeting of moving liquid masses of diverse origin. It is thus that on the surface of a pond, the waves that have risen at different points form an immense network of intersecting lines, which the breeze mingles in undecided wavelets.

* See below, p. 123.

CHAPTER XIV.

TIDAL CURRENTS.— RACES AND WHIRLPOOLS.—TIDAL EDDIES.— RIVER TIDES.

The popular belief is that the oscillations of the tides are always accompanied by currents changing regularly with the ebb and flow, and tending alternately in one direction or the other. This is, it is true, a pretty frequent phenomenon, especially at the mouths of rivers. Usually when the water rises, a tidal current rushes at the same time towards the shore and into the estuaries of rivers; then when the level of the liquid mass falls, a return or low-water current, swelled by the fresh water from inland, flows again towards the open sea. Nevertheless, this coincidence of the horizontal currents with the vertical oscillations of the ocean is far from being reproduced with regularity in all parts. The tide, being merely a swelling of the sea, can rise without the least movement occurring in one direction or the other. A remarkable example of this is seen in the Irish Sea, so rich in maritime phenomena. In the middle of the channel which separates the Isle of Man from Ireland, the sheet of water keeps perfectly tranquil between the contrary currents, though the water at this place rises more than 18 feet during the spring tides. On the other hand, as one can see at Courtown, on the coast of Arklow, the current determined by the meeting of opposing tides can have a great speed where the surface of the sea neither rises nor falls.* Finally, the same wave can follow a constant direction across two contiguous regions of the sea, one of which is at ebb and the other at flow.

The currents which occur in straits in consequence of differences of level, are sometimes extremely violent; and by their abrupt changes, their eddies and whirlpools may be classed among the most dangerous phenomena of the ocean. Thus the entrance to the Gulf of Normandy and the Channel Islands is rightly dreaded by navigators because of the terrible speed which the tidal currents attain there. The Blanchard Race, a strait which separates the Cape of La Hogue from the island of Alderney, is the first of these terrible marine defiles where

* See above, p. 113.

the ebb and flow, restrained between chains of rocks and shallows, move at the time of high water with a speed of nearly 10 miles per hour. Then comes the strait which bears the significant name of the Déroute Passage, and in which the currents flowing along the

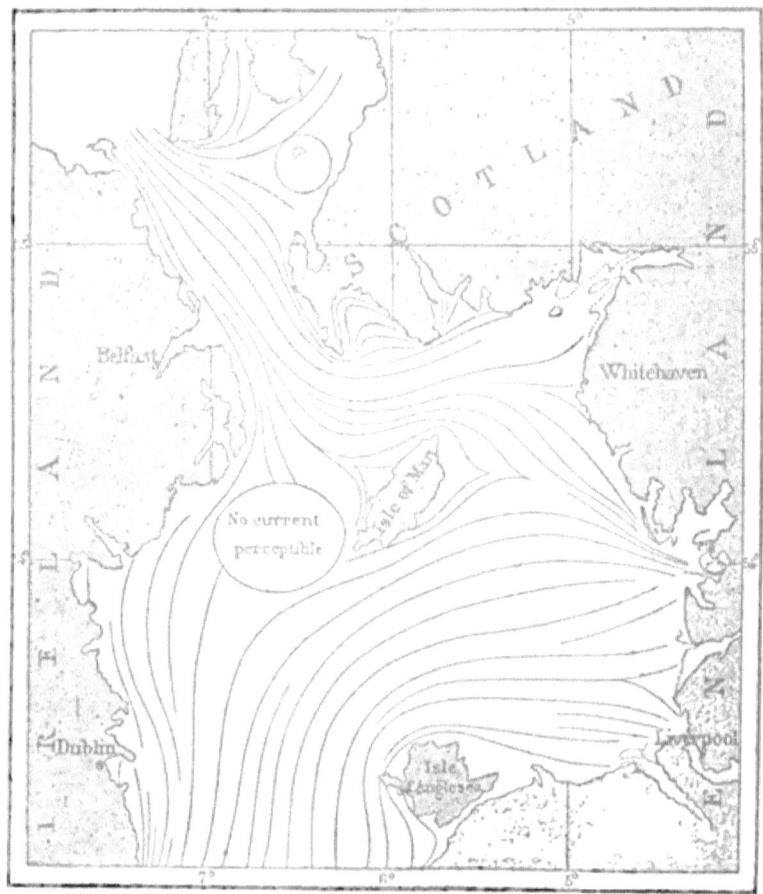

Fig. 41.—Course of the Tide in the Irish Sea.

rugged western coast of Cotentin meet those which come directly from the open sea by the breach opened between the islands of Jersey and Guernsey; there the marine rivers, less rapid, are nevertheless animated by a speed of nearly 10 feet per second.* Since the disaster of La Hogue, where Tourville, unable to sail against the for-

* Mounier, *Mémoire sur les Courants de la Manche.*

midable current of Blanchard Race, lost so many of his ships, how many vessels have been wrecked, how many crews have perished, in these terrible straits, which Victor Hugo has chosen as the theatre for his gloomy drama of *The Toilers of the Sea*.

The marine defiles which separate the British Isles from the continent, and especially those of the Hebrides, the Orkney, the Shetland, Färoe, and Lofoten Islands (whose rocks and shelving-banks confusedly stud a very uneven sea-bed, full of abysses), are also traversed by alternate tidal currents all the more rapid and tumultuous, because of the difference of level between the two sheets of water, which meet in the strait. The most formidable of these passages is perhaps the Great Gulf, or "Coirebhreacain,"* between the islands of Jura and Scarba, on the western coast of Scotland. At each change in the tide a current, flowing alternately towards the mainland and towards the open sea, is produced. The English Admiralty chart estimates its speed at nearly 11 miles per hour, but sailors affirm that it is at least nearly $12\frac{1}{2}$ miles, that is to say, more rapid than the stream of any continental river. No vessel can venture, in strong tides, into such a terrible race; especially when the wind blows in the contrary direction to the tide, for the Coirebhreacain is then in its entire extent a foaming "cauldron" without any visible limits.†

Other tidal conflicts are hardly less terrible; such for example is that observed in the straits of the Pentland Firth, between Scotland and the Orkneys, and which ends in the formation of currents estimated at more than 10 miles per hour. But the most celebrated of all these encounters between two tides of different levels is the Mosköestrom, towards the southerly extremity of the archipelago of the Lofoten Islands, called also by seamen the Maelstrom. The sombre imagination of northern peoples, always tending to the creation of monsters, saw in the strait of the Mosköe-strom a polype with arms several hundred yards in length, which caused the waters to whirl in an immense eddy, in order to draw ships into it and engulf them. From this ancient legend there has even remained with many the idea that this current is a sort of abyss in the form of a funnel, which floating objects approach by degrees, forming narrower and narrower circles, till they finally plunge for ever into this revolving well. But it is nothing of the sort. The only eddies are small lateral ones, produced by the meeting of the currents, and hardly 2 or 3 yards deep. The principal phenomenon consists, as in the Coirebhreacain and

* Gaelic, "Cauldron of the spotted seas."
† *Athenæum*, Aug. 26, 1864; *Mittheilungen von Petermann*, t. ix. 1864.

the Blanchard Race, of a rapid movement of the waters tending alternately in one or the other direction, at the time of the change of the tides. When in the open sea the flow rises in the direction from south to north, a part of its mass spreads with force into the strait opening to the south, between the two islands of Mosköe and Mosköe-naes. In proportion as the surface approaches a state of equilibrium, the current, gradually weakened, tends towards the south-west and then to the west. A period of calm follows these different movements of the waves, when the level is perfectly established; but soon the ebb commences, and tends in an inverse direction, at first towards the north, then towards the north-east and east. Thus in the space of one tide the waters are alternately carried, though with varying force, towards all the points of the compass.

The tidal currents, which occur at the entrance to rivers, frequently give place to tumultuous movements less terrible, it is true, than those of the races in archipelagos; but sometimes of an equally striking aspect. These phenomena are known under the name of the "bore," *barre*, "eager," or *mascaret*.

In penetrating into the estuary of a river, the tidal wave, retarded by the shallows and narrowed by its banks, must necessarily swell because of the restriction of the liquid mass in its bed. All the inlets and bays into which the tide penetrates present thus the spectacle of the "bore;" but in many passages the regular inclination of the bed, the uniformity of the shores, or else an intersection of various currents, diminish the first undulation of the tidal wave, or permit it to be confused with other irregularities of the surface. Elsewhere, on the contrary, all the topographical conditions are found united to give a great height to the "bore," and it then rises like a moving wall from one shore to the other of the estuary. At the mouths of certain rivers, such as the Amazon, the Hooghly, the Seine, the Dordogne, the Elbe, and the Weser, the waves of the "bore" assume enormous proportions at the time of high tides, and become formidable phenomena. In the Amazon, the "bore," called *pororoca* because of the roaring of its waters, rises, it is said, in three successive waves, attaining together from 30 to 50 feet in height; and vessels, surprised by this sudden flood, are in great risk of capsizing as in the open sea.

At the mouth of the Ganges the "bore" is also very formidable. As the old Hindoo legend says, in symbolic language, Bagharata having taken the divine Ganga as his spouse in the midst of snows, raised her in his arms, and, mounting his chariot, traced with its two large wheels the banks of the wide bed of the goddess. But when they arrived at the sea-shore Ganga recoiled with affright before the impure and monstrous ocean; she fled abruptly by a thousand channels, and since that epoch she comes and goes by turns, now venturing to descend, and now fleeing again towards the mountains, twice a day.*

It is in the bay of the Seine that the *mascaret*, or "eager," has been most regularly and carefully observed. Flowing from the open sea with a speed of from 15 to 20 feet per second, the liquid wall remains curved towards the centre, under the pressure of the fluvial current. The two points of the enormous crescent break in foam on the shores; while in the middle of the concavity, the even, rounded wave advances without even rippling the water before it. It seems to turn on the river like a gigantic serpent; rising from 6½ to 10 feet above the liquid plain; whilst behind it rise waves or *éteules* in concentric undulations quite as high, the advanced guard of the tidal mass. All the obstacles placed in the way of the *mascaret*

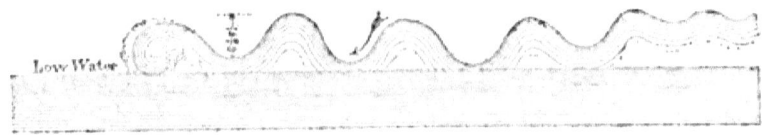

Fig. 42.—Profile of a tidal wave observed in the bay of the Seine (after M. Partiot).

irritate it by increasing its impetus; at length the tide, entering a

Fig. 43.—Height of the "mascaret" or tidal wave observed between Caudebec and Meilleraye (after M. Partiot).

wider and deeper part of the bed, gradually calms and moderates its height till it meets with another shallow or promontory. Moreover, each tide-wave is distinguished from the preceding by reason of the difference of winds, currents, and the masses of water put in motion. There is nothing more curious than to see, from the height of a promontory, two waves repelled obliquely by the banks crossing their furrows, and their *éteules*.

The sole means of diminishing the force of the *mascaret*, which in several estuaries, and especially in the bay of the Seine, is sometimes dangerous to small vessels, is to regulate the channel by deepening the shallows and straightening the banks. The works,

* Carl Ritter; Von Hoff, *Veränderungen der Erdoberfläche*, t. i. p. 378.

which insure a freer and deeper channel for navigation, are those which prevent the injuries caused by the great violence of the

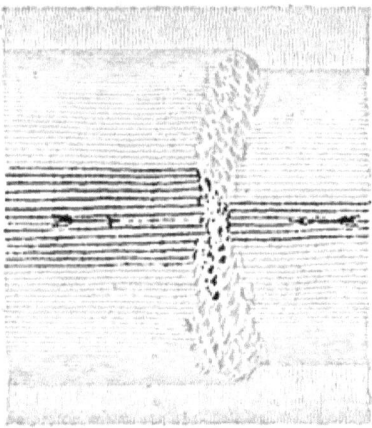

Fig. 44.—Plan of the "mascaret" or tidal wave observed in the narrows of the Seine (after M. Partiot).

tidal waves.* The *mascaret* of the Seine disappeared recently for some years, owing to the elevation of a bank of sand like a dike which prevented the entrance of the tide into the bed of the river. The encounter of the *mascaret* and the fluvial current have again

Fig. 45.—Plan of two tidal waves crossing each other's course on the banks of the bay of the Seine (after M. Partiot).

raised this bank of sand at a little distance. On striking against this new obstacle, the tidal wave rises up to surmount it. Different hydraulic works, undertaken in the beds of the Garonne and the Dordogne above the Bec-d'Ambez, have also often modified the phenomena of the *mascaret* there.

The sudden appearance of the tide in estuaries raises the fluvial waters very rapidly from the level of low to that of high water. At

* Partiot, *Annales des Ponts et Chaussées*, t. i. 1861.

Tancarville, which is the precise spot where the Seine discharges itself into the bay, and where the tide exceeds a mean amplitude of about 13 feet, the entire rising of the waters is accomplished in two hours, while the fall of the liquid mass, driven back by the tide, occupies about 10 hours. The river having to discharge during the period of ebb not only that which the flow had brought to it, but also the fresh waters from higher up, must follow its normal course towards the sea during a space of time longer than that in which it is driven back by the rising tide. For each point of the river-bed the duration of the flow is generally the shorter the farther that point is from the sea: the force of the tide is gradually exhausted, and towards the end of its course it only momentarily retards the speed of the fluvial current.

The amplitude of the tides diminishes, likewise, in proportion to their progress up the stream in rivers. The mass of fresh water

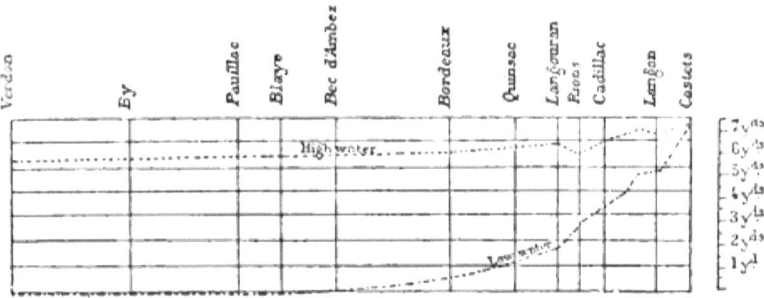

Fig. 46.—Tides of the Garonne.

flowing incessantly within the channel prevents the low tide from sinking, as it does on the sea-shore; and as to the high tide, its shorter duration does not allow it to rise to a much higher level than that which it attains on the strands and cliffs by the ocean. Thus, in the Garonne, the difference between the ebb and flow diminishes gradually above the Bec-d'Ambez, and near Castets, at about 95 miles from the sea, it is finally reduced to zero. In certain places, it is true, particular circumstances may cause apparent exceptions to this general law; a promontory rising before the tidal wave like that of Tancarville in the bay of the Seine, bars the way to the marine waters, and gives them in consequence a greater relative height above low water. But in spite of these abrupt projections, the mean amplitude of the tide diminishes from the lower to the upper course, and finally it becomes imperceptible.

CHAPTER XV.

EBB AND FLOW IN LAKES AND INLAND SEAS.—CURRENTS OF THE EURIPUS.—SCYLLA AND CHARYBDIS.

THE attraction of the sun and moon act no less on enclosed seas than on the great ocean; but in basins of small extent, the tide has not the necessary space to rise, and develop itself in an appreciable manner. Lake Michigan, which, although not less than 56,000 square miles in extent, is the smallest surface we are acquainted with where the regular return of the ebb and flow have been established with precision; the amplitude of the tide there, is, according to Lieut. Graham, less than 3 inches. Still, it is undoubted that the smaller lake-basins also experience normal oscillations every 12 hours; measures carefully made will probably reveal them one day.

Even in the vast Mediterranean the tides are very little perceived, excepting in the Gulfs of Syrtes, between the ancient Pentapolis and Tunis. In this part the phenomenon of the ebb and flow occurs with the greatest regularity, and one can study its progress as in the ocean. At the mouth of Oued-Gabès, almost at the end of the lesser Syrtes, the water alternately rises and falls at least 6½ feet. More to the north, in the port of Sfax, the average difference between high and low water is about 5 feet, but at the epoch of the equinoxes this difference attains to nearly 8 feet. Finally, at the Island of Djerbah, the ancient island of the Lotophagi, the mean amplitude of the tide is not less than 9 feet 10 inches.* This remarkable height of the tide on the shores of the Syrtes doubtless arises from the Mediterranean presenting in its southern part, from Port Said to Ceuta, a single basin, with a slightly sinuous bank, while on the coast of Europe it is divided into a number of smaller seas, those of Sardinia, the Adriatic Gulf, the Ionian Sea, and the Archipelago. Besides, the winds being much more regular on the African coast, the alternate play of the tides is not disturbed there, as on the coasts of Europe, which belong to the zone of variable winds.

* Victor Guerin, *Voyage Archéologique en Tunisie*, t. 1st.

However, an attentive examination of the movement of the waves has equally revealed to observers the existence of the tidal wave in the partial basins of the northern shores of the Mediterranean. Beyond Malaga, where the tides of the Atlantic are still propagated, the level of the sea hardly changes: but on the coasts of Italy the oscillations begin to be perceptible again. At Leghorn, the tide rises less than 12 inches; at Venice, the difference between the high and low waters varies from 1 to 3 feet.* At the mouths of the Po the tide does not attain the same height. On the coasts of Zante, in the Ionian Sea, it is less than 6 inches; finally, at Corfu, it does not exceed an inch.† In the Oriental basin of the Mediterranean, the tide is likewise very slight; nevertheless, the alternate oscillation of the sea is not ignored by the people living on the shores. Omar spoke doubtless of the tide when he said, 'The sea stands very high, and day and night it entreats the permission of God to inundate the land.'

Not only has the Mediterranean its ebb and flow like the ocean, but it has also its currents and eddies, and among these phenomena, there are some which, without being as formidable as the Mosköestrom or Blanchard Race, are not less celebrated, because of the glory with which classical antiquity has invested them. Thus, the Euripus, or Strait of Egripos, which separates the Island of Negropont from continental Greece, is said to be traversed by extraordinary currents, which produce with regularity their surprising phenomena. Up to the eighth day of the lunar month, the ebb and flow, whose mean amplitude is less than a foot, follow one another in a normal manner, only with one hour's delay; but from the ninth to the thirteenth day the movement of oscillation is suddenly hastened, and during the 24 hours no less than 12, 13, or 14 tides may be counted, each one having its flow, its period of stability, and its ebb. From the fourteenth to the twentieth day, a normal state of things prevails; then from the twenty-first to the twenty-sixth, every day will again be marked by a series of a dozen high and low tides. Such is the result of the experiences of the millers, who see the wheels of their mills turn alternately one way and the other, according to the direction of the current.‡ On their side, the Mussulmans maintain, as an article of faith, that the five waves of the

---

* G. Collegno, *Geologia dell' Italia*, p. 280.
† Von Hoff, *Veränderungen der Erdoberfläch. e*, t. iii. p. 256.
‡ Berghaus von Klöden, *Handbuch der Erdkunde*.

Euripus regularly follow the five hours of prayer;* finally, the rapid observations of several travellers describe in still another manner the oscillations of the sea in the narrow channel. The fact is, that the currents of the Strait of Negropont are unexplained, and if they succeed one another in as strange a manner as the inhabitants of those shores affirm, one would really comprehend the legend, according to which Aristotle, after having vainly sought to divine the mystery, plunged in despair into the whirlpools of the Euripus.

Still more famous than the currents of the Strait of Euboea were the abysses of Scylla and Charybdis, braved for the first time by the wise Ulysses. According to the Homeric chants, the two howling monsters which guarded the entrance to the Straits of Messina, drew into their submarine caverns immense whirlpools of water, which they afterwards discharged in furious currents, and all the ships which approached those formidable caverns were inevitably engulfed. At present there are no straits in the Mediterranean more frequented than those of Messina, and owing to the soundings effected in these pretended abysses where the ancients saw the navel of the sea, the monsters have lost their terrible prestige. It is now known that the whirlpools of Charybdis and Scylla are nothing else

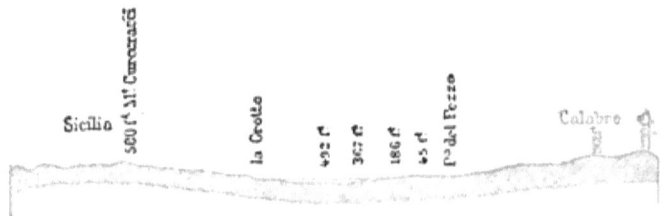

Fig. 47.—Profile of the Straits of Messina.

than lateral movements produced by the ebb and flow, in their passage through a too narrow channel, whose width is hardly 2 miles, and which the conquerors of Sicily have more than once crossed by swimming on their horses. At the time of the rising tide the current tends to the north from the Ionian to the Tyrrhenian sea; at the fall of the tide, the stream coming from the north assumes the preponderance, and drives the contrary current towards the south.† But there is a strife between the two liquid

* *Natur*, t. viii., 1864.
† Spallanzani; von Hoff; Smyth.

masses, and the field of battle moves incessantly from Messina to Scylla. On the confines of the currents, where the mingling of the waters is effected with violence, narrow eddies are formed, where the waves are more agitated than elsewhere; these are the 'cyclets,' or *garofali*. Ships avoid them, for fear of being too violently shaken; but they run no danger unless the wind blows strongly in a contrary direction to the tide. The strait is a curious spectacle, seen from the height of the mountains of Messina or Reggio, with the undulations and eddies that the conflicting waters describe; every instant sheets of water of a darker tint than those of the surface are seen to change their form, indicating the ebb and flow.

In the other enclosed seas of Europe the tides are likewise little felt. They are less than 16 inches on an average in the Zuyderzee, and during the days of the equinox or of tempests they hardly attain 3 feet 6 inches. The Baltic, which is much narrower and more strewn with islands than the Mediterranean, is subject in consequence to much slighter oscillations; it was even called in former times *morimarusa* (*mor y marb*), that is to say, in Celtic language, "Dead Sea."* The sailors pay no attention to the variations of the surface produced by the ebb and flow: for them the winds, the currents, and the meteorology of the atmosphere are the only phenomena which they have to observe. In fact, on the western coast of Jutland, the tide is on an average less than 12 inches, at the entrance to the Cattegat it loses still more in force and regularity, and in the straits of the Sound and the two Belts it is difficult to recognize. In the harbour of Copenhagen an oscillation of about 1 or 2 inches can still be sometimes distinguished, but only when the weather is perfectly calm and the surface of the water hardly rippled. At Wismar the phenomena of the tide are still more uncertain, and it is only by a series of observations on the surface of the waters pursued during several years, that the probable existence of a total variation of little more than 3 inches between high and low water can be ascertained. Near Stralsund the difference is only 1½ inch, and near Memel it hardly exceeds an inch. The much more considerable variations which occur in the level of the sea arise from the winds, the currents, or the pressure of the atmosphere. Rapid oscillations of nearly 3 feet have been sometimes seen to occur; but these are the *seiches*, similar to those of the lake of Genoa.† The force of the winds alone is sometimes sufficient to lower by little more than 3 feet

---

* Von Maack, *Geitschrift fur die Erdkunde*, 1860.
† See in Vol. I. the chapter entitled, *Lakes*.

the level of the sea in certain straits, as well as in the gulfs of Esthonia and Finland.*

The laws of the phenomena of the mouths of rivers differ entirely in the seas with strong tides, as the northern Atlantic, and in those with insensible oscillations, like the Baltic and the Mediterranean. In the estuaries where the sea rises regularly twice a day to a great height, it passes over every obstacle, bars, or sand-banks accumulated at the entrance to the mouths of rivers; while in those places where the level of the sea remains always the same, the dikes of mud or sand deposited parallel to the coasts between the fresh and salt waters, always close the entrance to the river. Thus the Rio Magdalena, and the Arato, in the Antilles; the Rhone, the Po, and the Nile in the Mediterranean, spread their liquid mass over bars which are often hardly a yard at the lowest part;† while the river of the Amazons, the St. Lawrence, the Gironde, and the Thames, allow free passage to ships at all hours.

This diversity of fluvial laws, according to the height of the oscillations of the tide, has the most important consequences for the commerce of regions watered by great rivers. In general the ports of the rivers without tide cannot be established at the mouth itself, because of the want of water, and merchants are obliged to choose a locality situated on the sea-coast at a certain distance from the sandy mouths of the river for their emporiums. Thus Marseilles, where almost all the commerce of the great basin of the Rhone is transacted, is constructed on the shores of a deep bay of the Mediterranean, far from the peninsulas of mud between which the river discharges itself. Alexandria, the great port of the Egyptian delta, lies to the west of the alluvial delta of the Nile; Venice is far from the mouths of the Po; Leghorn protects its port from the approach of the Arno; Barcelona is not at the entrance to the Ebro; and Carthagena in the West Indies and Santa Maria are only in communication with the great Magdalena by means of hardly navigable canals. The exceptions to this rule are not very numerous, still we may cite Dantzig on the Vistula, Stettin on the Oder, and Galatz on the Danube.‡

In seas with high tides the principal ports are found, on the contrary, not on the maritime coast-line, but on the rivers, and even at a certain distance from the mouth, not far from the place where

* Von Sass, *Bulletin de l'Academie de St. Petersbourg*, t. viii. 6.
† See in Vol. I. the chapter entitled, *Rivers*.
‡ Ernest Desjardins, *de l'embouchure du Rhone*.

the tide rises twice a day, thus changing the river into a true maritime gulf. London, Hamburg, Nantes, Bordeaux, Rouen, and many other great commercial cities, have been gradually built, in consequence of the necessities of commerce; as far as possible inland at the precise spot where the depth of water and the force of the tide allow ships to approach easily. Nevertheless, since the ships of the present day draw much more water than those of our ancestors, the result is that a number of ports on rivers have become insufficient. It is thus that Rouen has been gradually replaced by Havre as the port for international commerce. Thus Nantes, too, has seen in these days a rival city grow up in the village of St Nazaire, so modest but a few years ago. Perhaps the hamlet of Verdon, provided sooner or later with docks, basins, and jetties, will become likewise the real commercial Bordeaux.

BOOK IV.—THE SHORES AND ISLANDS.

CHAPTER XVI.

INCESSANT MODIFICATIONS OF THE COAST-LINE.—THE FJORDS OF SCANDINAVIA AND OTHER COUNTRIES NEAR THE POLES.

THE sea, every wave of which contains perhaps thousands of living organisms, seems itself to be animated by a vast and mighty life. Ever-changing hues, dark as fog or brilliant as the sun, pass over its immense extent, its surface ripples in long undulations, or rises in bristling waves; its shores are touched with a border of foam, or disappear under the white mass of breaking surf. Sometimes it breathes a scarcely audible murmur, and again it combines in very thunder the roarings of all its waves dashed and broken by the tempest. By turns it is smiling and terrible, gracious and formidable. Its aspect fascinates us; and as we walk along its shores, it is impossible to avoid contemplating and interrogating it ceaselessly. Ever moving, it symbolizes life, in distinction to the silent and passive earth which it assaults with its waves. And besides, is it not always untiringly at work to modify the contour of the continents, after having once formed them layer by layer in the depth of its waters?

The most important part of the geological labours of the ocean is hidden from our eyes; for it is at the bottom of its abysses that the sea deposits the silica, limestone, chalk, and conglomerates of every kind which will one day constitute new lands. But at least we can witness the continual modifications to which the incessant movement of the sea subjects the shores. These modifications are considerable, and during the historical ages a number of coasts have already completely changed their form and aspect. Promontories have been razed, while at other parts points have advanced into the waves; islands have been transformed into reefs; others have been entirely swallowed up; others again joined to the mainland. The sinuous

line of the shore has not ceased to oscillate, encroaching here on the
waters of the ocean, and there on the continental surface. The
action of the sea is double : it is constantly re-touching the contours
of its basin, either by wearing away the rocks that border it and
carrying away the strand, or by casting up on its coast the alluvium
and wreck of every kind that it tosses in its waves. All that it en-
gulfs on one side it gives back elsewhere under another form.

Before the sea had modified its shores by destroying peninsulas
and filling up bays and estuaries, the form of the coast was certainly much less regular than it is now in the outline of most countries.
If the marine waters were raised by a sudden revolution to 100 or 200
yards above their present level, the ocean, inundating all the river
valleys to a very great distance from the present shores, would suddenly enter in elongated gulfs into the depressions of the continent,
and change all the valleys and lateral gorges into bays. In the place
of each of those river-mouths which hardly indent the normal line of
the coast, deep hollows would be opened, dividing into numberless
ramifications. But a work in the opposite direction will instantly be
commenced when this change in the outline of the shores is accomplished. On the one side, the water-courses, bringing down their
alluvium, will gradually fill the upper valleys, and little by little
restrict the domain of the maritime conquests. On the other side,
the ocean will also labour by its dunes along the coast, its banks of
sand or shingle, to take away from its surface all those new bays that
the sudden increase of its waters had given it. After an indefinite
lapse of centuries, the shore would finally re-assume the gently un-
dulated form that the greater number of coasts now present.

There are still many countries where this double work of the sea
and the continental waters has hardly commenced. Those lands
whose coast-line, thus preserving its first form, is still deeply indented, are all situated at a great distance from the equator, in the
neighbourhood of the Polar zone. In Europe, the western coasts of
Scandinavia, from the promontory of Lindesnaes to the North Cape,
are jagged by a series of these *fjords*,* or ramified gulfs, and not
only the shore of the continent, but all those islands also which form
a sort of chain parallel to the Norwegian plateau, are fringed with
peninsulas and cut into by small fjords, winding in immense passages.
Among these indentations, which increase the length of the coast ten-
fold, and give to the coast-line a border of innumerable peninsulas,
more or less parallel, some are pretty uniform in aspect, and resemble

* Called in Scotland, *firths*.

enormous trenches, hollowed out in the thickness of the continent; others are divided into several lateral fjords, which make the inland waters an almost inextricable labyrinth of channels, straits, and bays. The total development of the coasts is so much increased by these indentations, that the western shore of the peninsula, whose length in a straight line is about 1180 miles, is increased to above 8000 miles by the bends and turnings of the shore, which is more than the distance from Paris to Japan.

The plateaux of Scandinavia, terminating abruptly above the North Sea, the slopes which command the sombre defiles of the fjords, are almost always very steep; there are some which rise in perpendicular or even overhanging walls, serving as a pedestal to high mountains. It is thus that the Thorsnuten, situated to the south of Bergen, on the shores of the Hardanger Fjord, attains an elevation of above 5250 feet at less than 2¼ miles from the shore. In many a bay of western Norway cascades are seen to leap from the top of the cliff, and precipitate themselves in one jet into the sea, so that vessels can glide between the walls of the rocks and the parabola of the roaring cataracts. Below the water the escarpements are continued also in most of the gulfs, so that in certain defiles of rocks, whose breadth from cliff to cliff is only from 300 to 600 feet, the lead must be

Fig. 48.—Lysefjord, Norway.

thrown to a depth of from 272 to 327 fathoms before touching the rocky bottom.* In the *Toilers of the Sea* Victor Hugo correctly cites the Lysefjord as most fearful to contemplate among its gloomy ap-

* Berghaus, *Was man von der Erde weiss*, p. 280.

proaches, many of which are for ever deprived of a ray of sun by the high walls of rock which enclose them. This enormous cutting, of an almost perfect regularity, penetrates above 26 miles into the interior of the continent, though in several places it hardly exceeds 1965 feet in breadth; its walls rise from 3270 to 3600 feet in height, and near the edge the lead only touches the ground at about 220 fathoms.* Doubtless the first seaman who sailed over the dark, tranquil waters

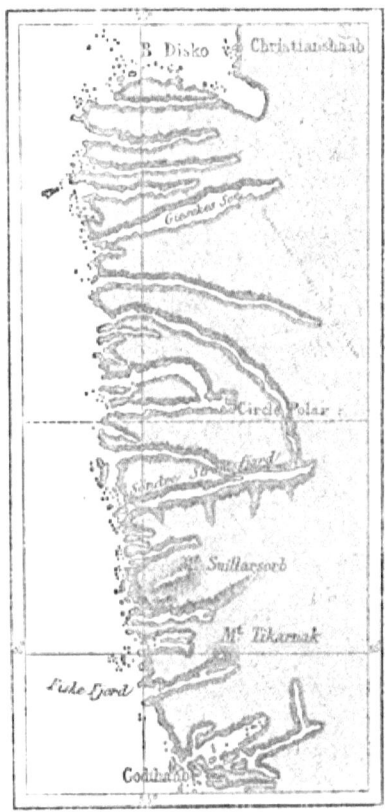

Fig. 49.—Fjords of Greenland.

of this abyss must have advanced with a sort of horror, asking at each new turn of the approach, whether he was not going to see some terrible god rise before him. Even now it is not without a

* Vibe, *Küsten und Meer Norwegens, Mittheilungen von Petermann*, 1860.

shudder that one penetrates into this gloomy defile, where the ancients would doubtless have seen the entrance to the infernal regions.

The islands of Spitzbergen, Faröe, and Shetland present also in their outline hundreds of fjords, like those of Scandinavia. The coasts of Iceland, Labrador, and western Greenland, those of the islands of the Polar Archipelago, and finally the American coast-line of the Pacific, from the long peninsula of Alaska to the labyrinth of Vancouver's Islands, are no less rich in indentations than the coast-line of Norway. The shores of Scotland are deeply cut in the same way, but only on the western side, where there are besides numerous islands reproducing in miniature the maze of promontories and bays of the mainland. That part of Ireland turned towards the open

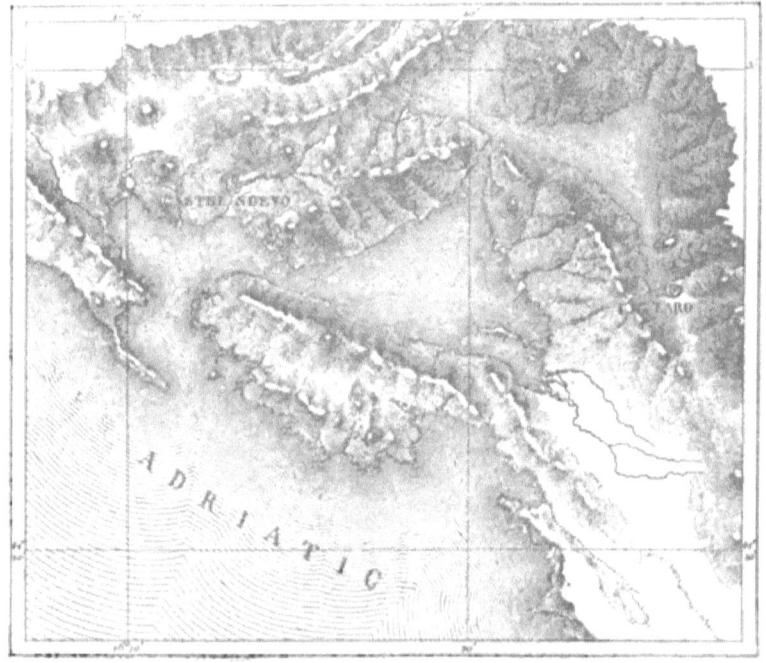

Fig. 50.—Mouths of Cattaro.

sea develops itself also into a succession of rocky peninsulas, separated by narrow gulfs; but to the south and east the coasts of the British

FJORDS OF SOUTH AMERICA.

Islands are much less varied in form, and sweep in long regular curves. In France we hardly find a vestige of indentations like those of the Norwegian fjords, except at the extremity of Brittany; and there does not even exist a word in the language to designate them. In Spain, in the same way, the part of the peninsula turned towards the north-west, and where the ports of Ferrol and Coruña

Fig. 51.—Fjords of South America.

open, is the only one which presents some lines of fjords half filled up. Two countries on the borders of the Mediterranean have their coasts also cut into fjords, partially obliterated by alluvium; these are Asia Minor and Dalmatia, whose high mountains, formerly covered with glaciers, overlook narrow bays with fantastic outlines like the mouths of Cattaro; but along these two shores the peninsulas of the coast-line are still uniformly turned towards the west.

To the south of the Adriatic and the Archipelago on the coast-line of warm or torrid countries no more fjords are seen. To find a similar formation of shores we must traverse the entire continent of America to its southern extremity. The fjords only commence beyond the uniform coast-line of Chili, with the Island of Chiloe, its numerous bays, and the network of straits in the Archipelago of Magellan and Terra del Fuego. This is the only region in the southern hemisphere where the astonishing phenomenon of tortuous and deep valleys filled by the waters of the sea is witnessed. As to the countries of the Antarctic continent, no indentations can be recognized in them, since the contours of the bays and capes, the gulfs and peninsulas, are all filled by the snouts of glaciers and by continuous ice-fields.

CHAPTER XVII.

FILLING UP OF THE FJORDS BY MARINE AND FLUVIAL ALLUVIUM.

THE comparative study of all the shores leads thus to the confirmation of this fact, that fjords are only met with on the coasts of cold countries, and that, with equality of temperature, they are much more numerous and better developed on the western coasts, than on those turned to the east. Why does this strange geographical contrast occur between the various shores according to the position which they occupy to the north or south, to the west or east? Why have the strands and even the cliffs, bathed by a warm or temperate atmosphere, assumed in the outline of their curves such a great regularity, while the valleys, opened in the thickness of the plateaux of Scandinavia, Greenland, and Patagonia, have preserved their primitive form? A cause whose effects are produced at the same time and in the same manner at the two extremities of the continents, in the northern islands of America and Europe and in the Magellanic Isles, must necessarily have been a great geological phenomenon, acting during an entire age of our planet.

This phenomenon was the special climate which during the glacial period made itself felt on the surface of the globe, and transformed the mountain snows into long rivers of ice. The map speaks for itself, so to say; it relates clearly how the fjords, those ancient indentations of the coast-line, have been maintained in their primitive state by the prolonged sojourn of glaciers.* In fact, the cold period, the unequivocal witnesses of which are still to be seen even in the tropics and under the equator, at the foot of the Andes and in the valley of the Amazon, naturally lasted longer in the vicinity of the poles than under the torrid zone and in the temperate regions. This glacial period, which terminated perhaps thousands of centuries ago on the burning shores of Brazil and Columbia, has ceased at a relatively recent epoch on the coasts of France and England. At an age still

---

* See in Vol. I. the chapter entitled, *Snow and Glaciers*—Oscar Peschel, *Ausland*, 1866.

nearer our historical time the fjords of Scandinavia have been in their turn freed from the glaciers that filled them, whilst quite in the extreme north and in the Antarctic regions there are countries where the rivers of ice still descend into the sea, and stretch far into the gulfs. The glacier of the Bay of Magdalene, which Messrs Martins and Bravais have explored, projects far into a fjord which is 55 fathoms deep, and the terminal cliff of ice, driven out by the weight of the upper snows, presents a curved line, turning its convexity towards the open sea. On still colder coasts, such as the north of Greenland, and at the South Pole, the outline of the Antarctic countries, even the bays are entirely filled up with ice, and this running into the sea gives a regular outline to the whole coast. The waves of the open sea dash against a long wall of crystal, and the icy layers disguise the true form of the architecture of the continents, as the fluvial alluvium and marine sandbanks do in other climates. Nevertheless deep valleys, hidden by the ice-fields, are also cut into the line of these Polar coasts too, and in a future geological period, when the ice shall have disappeared, these incisions of the continent will become in their turn fjords, similar to those of Scandinavia.

At the epoch when the bays of Scandinavia were filled with ice as those of northern Greenland are in our days, they preserved their primitive form, excepting that the lateral walls and the rocks at the bottom were grooved and polished by the friction of the mass in movement and the fragments which it carried with it. The blocks of stone fallen on the snow, and on the surface of the glacier, the heaps of pebbles and earth torn by storms and thaws from the sides of the mountain, formed moraines exactly similar to those which are now seen on the diminished glaciers of the Scandinavian mountains. But these moraines, instead of crumbling away with the ice, in some valleys thousands of feet above the sea, were carried to the very mouths of the fjords in the open sea, and plunged into the middle of the waves with the pieces detached from the glacier itself. The successive débris of rocks and pebbles must necessarily gradually raise the frontal submarine moraine, and in fact at the entrance of all the Scandinavian fjords, heaps of deposit are found rising like ramparts out of the deep water. The seamen of Norway give the name of "sea gates" to these natural barricades, which serve as limit to the ancient glaciers, and where the fish from the neighbouring waters assemble in myriads. Off the coasts of western Scotland, as at the entrance to the small gulfs of Finisterre, the ridges of submarine banks and reefs are observed, which are probably nothing else than ancient terminal glacial moraines.

After the period which preceded the present era, the glaciers of Scandinavia retreated little by little into the interior of the fjords, then ceased to touch the level of the sea, and their lower extremity mounted higher and higher in the open valleys on the sides of the mountains. It was then that the immense geological labour of filling up the bays commenced for the torrents and the sea. The fluvial waters brought their alluvium, and deposited it as an even strand at the foot of the mountains, while the sea levelled with sheets of sand or mud all the fragments of rocks which it had worn away by its waves. Already in a great number of Norwegian fjords this work of transforming the domain of the waters into firm land has made very sensible progress, and if we knew the amount per century of the augmentation to the shores, we should be able to calculate approximately the epoch at which the valley was free from ice. On the inclined eastern side, towards the open country of Sweden, an analogous work is accomplished; there, the glaciers have been replaced, not by the waves of the sea, but by the lacustrine waters divided into different basins, and these waters also retreat gradually before the alluvium of the torrents. In the same way, in the great chain of the Swiss Alps, several deep depressions, formerly the beds of immense glaciers, have become a sort of continental fjord, such as the lakes of Maggiore, Iseo, Lugano, Como, and Garda.* These lacustrine basins are closed at the south by large moraines, like the sea-gates of Norway, and their waters, like those of the fjords, are gradually displaced by the alluvium brought down by Alpine torrents.

Situated more to the south than the fjords of Scandinavia, and nearer the source of the warm current flowing from the Antilles, the western bays of Scotland must have been free from ice long before the coasts of Norway, and it was still earlier that the indentations of the coast-lines of Ireland and Brittany ceased to serve as beds to the solidified snows of the surrounding mountains. As to the shores of the British Islands turned to the east towards the North Sea, they have certainly long been freed from ice, for then as now the winds from the west and south-west prevailed in Europe, and carried the rains over the slopes of the mountains inclined towards the Atlantic; on the opposite slope the glaciers are sooner melted, because of the want of the necessary moisture. This is the reason of the striking contrast presented in the British Isles and

* Oscar Peschel, *Ausland*, 1866.

Iceland by the western coasts all cut into deep bays, and the eastern shores whose fjords are less deep, or even completely obliterated by

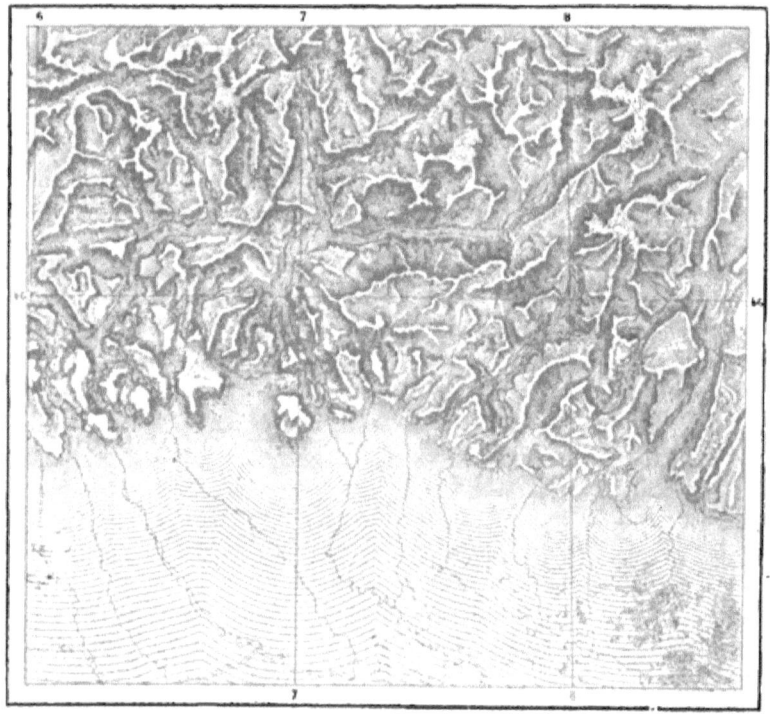

Fig. 52.—Ancient Fjords of Northern Italy.

alluvium. In the same way, at the south of America, the rains being much more abundant on the western slope of the mountains of Patagonia, the glaciers have descended much lower into the valleys, and the fjords, preserved by the ice in their primitive state, make all this part of the American coast-line a real labyrinth. The form of the continents themselves must be explained by the movements of the atmosphere.

After the retreat of the glaciers, the work of rendering the shores regular goes on in the various countries with more or less rapidity, according to the form of the continents, the depth of the fjords, and all the phenomena which constitute their geographical circumstances. In certain countries where the rivers are of little importance, as in the peninsulas of Denmark and in Mecklenburg, the fjords are first closed on the seaward side, and then become long and narrow lagunes, separated from the salt waves by the sandy beaches. Those gulfs, on the

Fig. 53.—Fjords of the South-East of Iceland.

contrary, where great rivers discharge themselves, are gradually filled up by alluvium in those parts the furthest from the ocean, and are changed little by little into estuaries. Finally, many shores, among others those of eastern Iceland, present a great number of fjords, one beside the other, which are narrowed at the same time above and below by the deposit from the sea, and that of the streams from the interior. It is thus that a multitude of ancient gulfs in Scandinavia, England, and France, have been gradually changed into dry land. The gulfs of Christiansand in Norway, of Carentan in France, formerly projected in all directions from deep abysses, the place of which is occupied now by fields and marshes.

Whatever may be the diversity of means employed by nature in filling up the ancient glacial bays, the labour is accomplished in due time, and we may state that from the temperate to the equatorial zone the curves of the shore increase in regularity. The innumerable ports which penetrate deep into the northern lands, are suc-

Fig. 54.—Filled-up Fjords of Christianssand.

ceeded in the south by more and more inhospitable maritime shores, because of destitute indentations. And on the coasts of the torrid zone, which are destitute of the mouths of rivers, vessels must sail along for hundreds of leagues before finding a harbour of refuge. It is the three southern continents, South America, Africa, and Australia, which present in their outline a most uniform development of coast and are most destitute of bays.

If we can rightly consider each glacier as a natural thermometer, indicating by its advance and retreat all the changes of local temperature, we may in the same way regard the general character of the coasts, from the fjords of Greenland and Norway to the long shores of equatorial Africa, as a visible representation of the changes of temperature which have taken place on the surface of the globe since the glacial epoch. If by long and patient study we succeed in measuring the time which is necessary for the alluvium of the sea and rivers thus to modify the forms of valleys once filled with

TIME MEASURED BY ALLUVIUM. 113

ice, we can then estimate the amount of time which has elapsed since the glacial epoch. This vague period, which according to various geologists comprehends thousands or millions of years, will

Fig. 55.—Ancient Fjords of Carentan.

assume, at least for the times nearest to us, a more precise meaning, and will arrange itself like the centuries in the chronology of mankind.

CHAPTER XVIII.

DESTRUCTION OF CLIFFS.—THE COASTS OF THE CHANNEL.—THE STRAITS OF DOVER.—ACTION OF SHINGLE AND SAND.—GIANTS' CAULDRONS.—SPOUTING WELLS ON THE COASTS.—TIDAL WELLS.

ALTHOUGH there is necessarily an equilibrium between the work of demolition and that of reconstruction, we would nevertheless at first sight be tempted to believe that the sea took the greatest pleasure in destruction. On contemplating the cliffs, those perpendicular walls which on various coasts rise many hundreds of yards above the level of the sea, we are struck with awe to see how the repeated assaults of the waves have been sufficient thus to cut the mountains and hills whose bases were formerly gently sloped to the water. From the top of these cliffs, we see the tumultuous ocean spread at their feet like a plane surface, and we no longer distinguish the billows but by their reflections, or the breakers but by their garland of foam; the multiplied sound of the waves melts into one long murmur, which dies away and rises to die away again. And yet this water, which we see below at such a great depth, and which seems powerless against the solid rock, has thrown down piece by piece all that part of the hill or mountain, of which the cliff is but a gigantic memorial: then, after having thrown down these enormous masses, it has reduced them to sand, and perhaps caused the very trace of them to disappear. Often not even a rock remains where promontories once jutted out. The phenomena ascertained even during the short life of man, are facts so grand in their progress, and so remarkable in their effects, that an English savant, Captain Saxby, has proposed to make of them a special science, *Ondavorology*.[*]

To gain some idea of the destructive force exercised by the waves of the ocean, it is sufficient to contemplate them on a tempestuous day from the height of the chalky cliffs of Dieppe or Havre. At our feet we see the army of whitening billows rush to the assault of the rocks. Driven at the same time by the wind, the tide, and the lateral current, they leap over the rocks and shelves of the shore,

[*] *Nautical Magazine*, Jan. 1864.

ENCROACHMENT OF THE SEA. 145

and strike the base of the cliffs obliquely. Their shock causes the

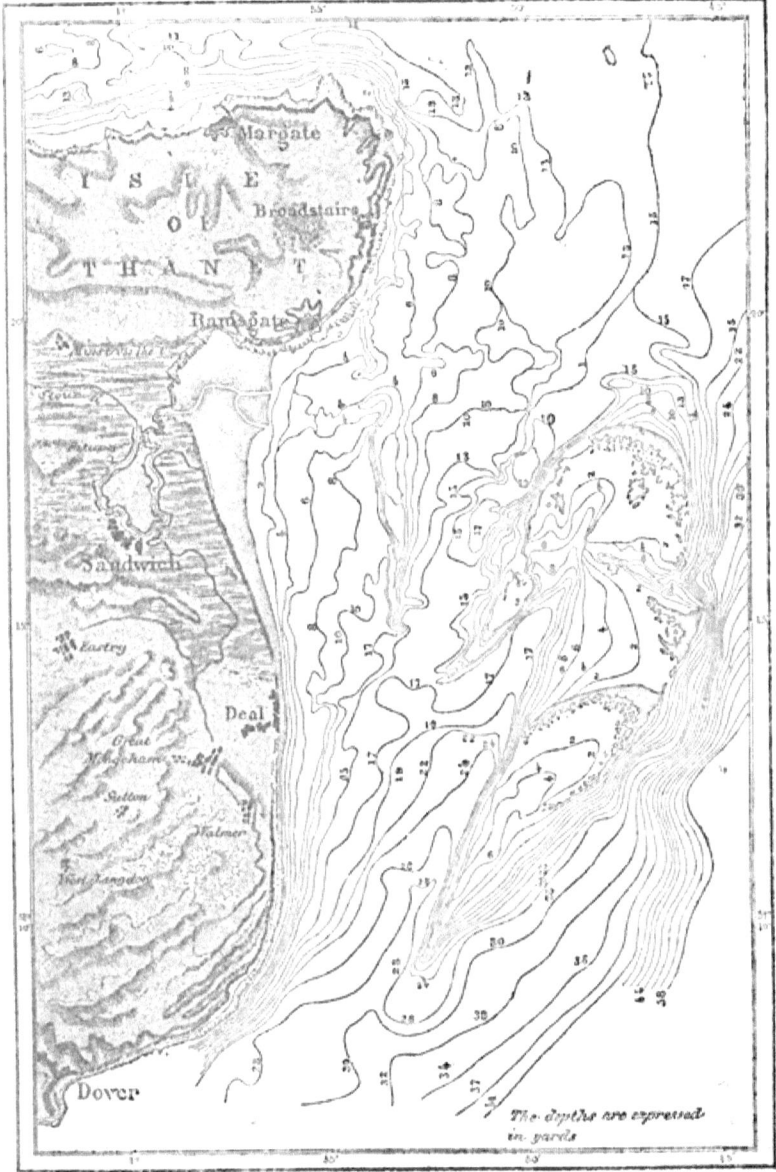

Fig. 56.—Roads of the Downs.

enormous walls to tremble to the very summit, and the roar reverberates in all their angles with an incessant thunder. Dashed into the fissures of the rock with terrible force, the water sweeps away all the clayey and chalky matter, and gradually lays bare the solid beds, wrenches large blocks out of them, rolls them on the strand, and breaks them into shingle, which it drives along with a dreadful noise. Through the eddy of boiling foam which besieges the shore, one can only now and then perceive the work of demolition; but the waves are so laden with fragments that they present a blackish or earthy colour, as far as the eye can reach.

When the storm has ceased, the encroachments of the sea can be measured, and we can calculate the millions of cubic yards of stone engulfed or transformed into shingle and sand. Towards the end of the year 1862, during one of the most terrible tempests of the century, M. Lennier saw the sea batter down the rocks of La Hève to a thickness of more than 50 feet. Since the year 1100, the waters of the Channel, aided by rain, frost, and other agents, that act strongly on the upper strata, have cut down this cliff by more than 1500 yards, that is to say, more than two yards per year. The spot where the village of Sainte Adresse formerly stood, has given way before the flood, and is replaced by the bank of l'Eclat.* M. Bouniceau, one of those savants who have specially studied the phenomena of erosion of shores, estimates the fraction of cliff which is carried away by the sea on the coasts of Calvados at above a quarter of a yard on an average yearly, while on the coasts of Seine Inferieure the annual erosion may be considered as nearly a foot.

In some places on the southern and eastern coasts of England, the invasions of the sea take place with an equal or even superior rapidity, for the farmers generally count on the loss of about a yard per year along the cliff.† To the east of the peninsula of Kent, the waters have advanced more than 3 miles towards the west since the Roman period. In their successive invasions, they have submerged the vast domains of the Saxon Earl Goodwin, and have replaced them by the terrible Goodwin Sands, where so many ships are lost every year: and they have transformed the narrow lagune of the Downs into great open roads. According to the calculations of M. Marchal,‡ the total amount of denudation which the waters of the eastern part of the Channel carry on every year is equal to above 13 millions of cubic yards.

* Lamblardie, Bande, *Revue des deux Mondes.*
† Beete Jukes, *School Manual of Geology*, p. 90.
‡ *Annales des Ponts et Chaussées*, 1er sem. p. 201.

The Straits of Dover are being continually enlarged by the action of atmospheric influences, the waves, and the current which flows from the Channel into the North Sea. The patient researches of M. Thomé de Gamond, an engineer to whom we owe the fine project of the international tunnel between France and England, have proved that the cliff of Gris-Nez, the nearest point of the French coast to Great Britain, loses on an average more than 27 yards per century. If in former ages the progress of erosion was not more rapid, it would be about 60,000 years before the present epoch that the isthmus connecting England with the continent was broken by the pressure of the waves. Nevertheless, it is impossible to indicate any date, since at this place the ground has sunk and risen at various intervals; ancient beaches four or five yards above the present level of the sea, as well as submerged forests, testify to these successive oscillations.*

Along the coast of France, to the east of Cape Antifer, the pebbles resulting from the denudation of the cliffs are continually advancing towards the mouth of the Somme. Arrested at about 6 miles beyond these last flinty cliffs, by the promontory of Hourdel, so named from the dash (*heurt*, Fr.) of the waves, they are subsequently taken up by the current which runs towards the Strait. Triturated more and more, they travel from sand-bank to sand-bank, and after having passed the Strait, are deposited in beds of mud either on the surface of the innumerable banks of the North Sea, or on the coasts of Flanders, Holland, and eastern England. It is these deposits, which are called by the expressive name of *gain de flot* ('winnings from the waves') in the neighbourhood of the Channel. The 10 millions of cubic yards of fragments taken annually from the cliffs of Sussex and Kent, as well as from those of Calvados and the Pays de Caux, are carried back to the coasts of the northern countries, and it is at the expense of the shores of the Channel that the *polders* of Holland and the fens of Norfolk and Lincolnshire are formed. In consequence of this double work of erosion at one point and deposit on another, the shores situated to the north of the Straits present a perfect contrast with the coasts of the Channel. While the cliffs of France and England, on the borders of this sea, are cut into concave bays, the beaches which stretch to the north of the Straits of Dover uniformly exhibit a convex arrangement. The waves give back in sand and mud what they have taken in rocks and boulders.†

* Day, *Geological Magazine*, 1866.
† Marchal, *Annales des Ponts et Chaussées*, 1er sem. p. 204.

We must not think that it is the force alone of the breakers that demolishes the cliffs along the shore. The sea would be almost powerless against the hard rocks, if, on approaching the shore, it was not charged with all kinds of *débris*, blocks and pebbles, sand and shells, projectiles which are hurled by every wave against the cliffs which oppose them. Using thus the stones that have fallen as so many battering-rams, the billows roll them over the strand to the foot of the cliffs, dash them against the projecting points, and finally break

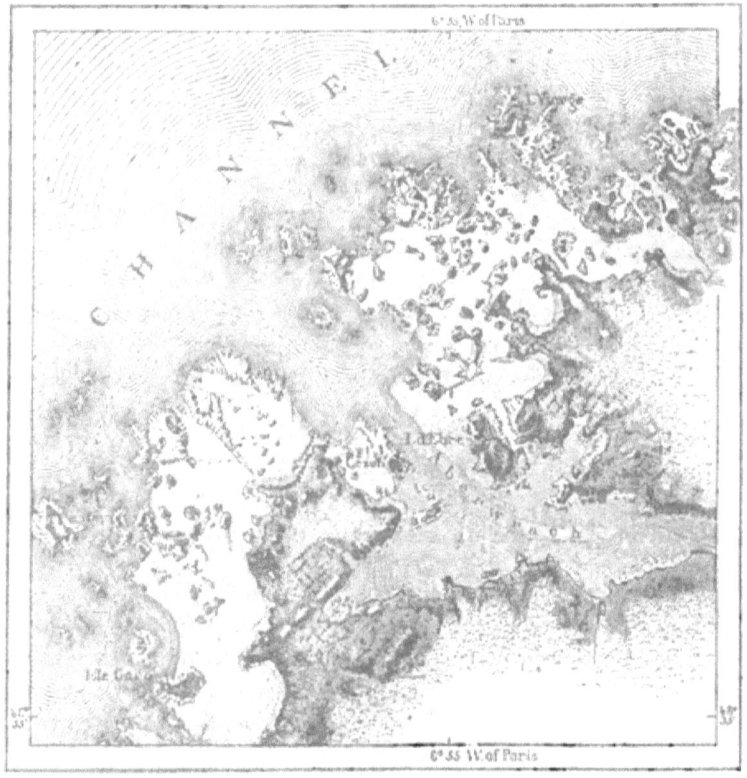

Fig. 57.—Map of Abervrac'h.

off masses and reduce them to sand. The sand itself, incessantly washed against the rocks, wears away the most solid layers little by little, and thus continues the work of destruction commenced by the shingle; it is in great part the fragments of the promontory itself which serve to further its destruction. On all the rocky coasts of Scandinavia, Scotland, Ireland, and Brittany, the multitude of reefs

THE GIANTS' CAULDRONS.

that extend seawards for a great distance from the shore are nothing else than the ancient foundations of the continent, which have been gradually razed by denudation to a level with the water. From the top of any hill on the coasts of Paimpol, Morlaix, and Abervrac'h, we may thus distinguish at low tide what was the primitive form of the shore.

The deep and regular excavations known under the name of "giants' cauldrons," are the most curious of the geological feats accomplished by the scattered blocks. Every stone reposing on a ledge of the rock where the waves break, hollows out during the course of ages a kind of well, the walls of which are polished, and planed by the friction. Finally, these cavities, where the gradually rounded

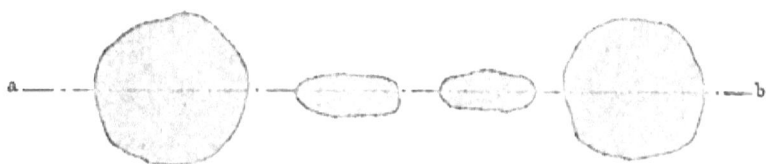

Fig. 58.—"Giants' cauldrons" of Haelstolmen.

stone does not cease to oscillate, acquire a depth and width of several yards, and these are then, according to tradition, the cauldrons where

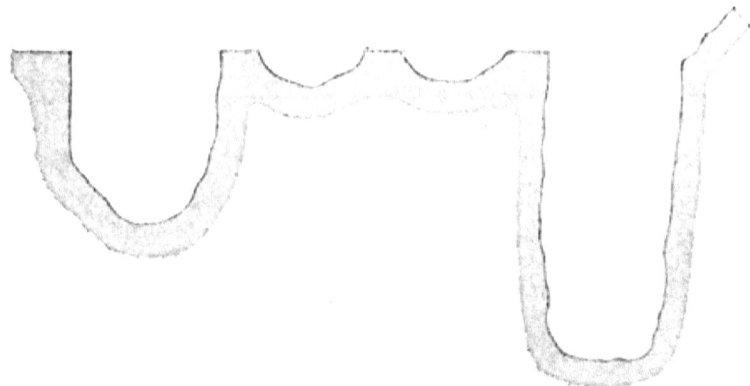

Fig. 59.—Section of the "Giants' cauldrons" of Haelstolmen, taken along the line *a b* in Fig. 58.

the giants prepared their repasts in former times. Very remarkable excavations of this kind exist on the coasts of Scandinavia, where blocks of granite, rolled along by a furious sea, are retained by abrupt rocks in a great number of cavities.

A phenomenon not less interesting than the revolving of the stones in the giants' cauldrons, is the sudden appearance of columns of sea-water, which spring in jets through the fissures of the rock. When a large wave is swallowed up in one of the fissured caverns on the coast, its force is sometimes so great that the rock resounds as with the discharge of artillery. The mass of water drives the air before it, and not finding in the walls that surround and compress it a large enough space to develop itself, springs through the crevices of the vault. Most of these fissures, gradually sculptured anew by the waters which escape from them, at length assume the appearance of real wells, where each return of the wave is signalized by a sort of *geyser* of variable dimensions. There are some which spring several yards high, and can be seen at a great distance, like the jet of water by which the whale betrays himself afar off; hence arises the name of blowers (*souffleurs*) given in many countries by sailors to these phenomena on the shore.

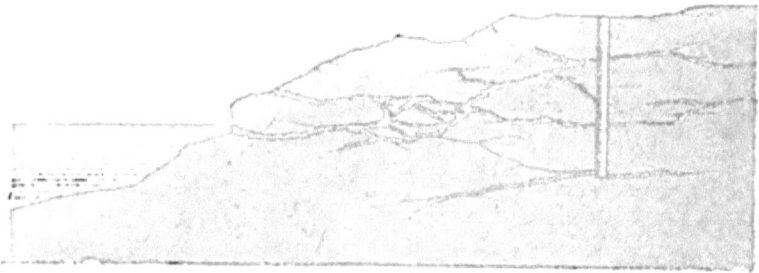

Fig. 60.—Tidal wells.

The pressure of the tide does not make itself less felt than the force of the waves in the interior of the fissured rocks of the coast. It does not, it is true, cause magnificent fountains to spring far above the sea, but it lowers the level of the water in all the wells near enough to the shore; even in those that are filled with fresh-water. And this is what theory could have indicated beforehand; the mass of water that penetrates far into the crevices of the rock, retains the waters which infiltrate from the interior, the latter, salt or fresh, remain in their reservoirs, and rise at the same time as the tide; then, when the ebb commences, they flow into the sea, and overflow again as soon as the pressure of the rising water ceases. Where the rocks of the coast are much fissured, which is almost everywhere the case with cliffs composed of calcareous strata, there exist these "tidal" wells, which rise and fall alternately with the tide. We may specially cite those of Finland, near Wasa, those in the environs of Royan, on the right bank of the Gironde, and, above all, those of the Bahama Islands. In many of these islands, all the wells, without exception, are regulated by the flow of the sea.*

There are even certain coasts opening so deeply into large hollows, on the side towards the sea, that the waves penetrate to a great distance into the interior of the continent. A curious example of this is seen in that part of Louisiana known under the name of the Attakapas. There, the prairies of the coast, protected against the tempests of the Gulf of Mexico by chains of sand-banks and long islands parallel to the shore, incessantly gain upon the ocean. But they are only solid on the surface, for their roots are bathed by the sea-water, which advances far into a bay with invisible outlines. The fishermen do not fear to venture on these floating meadows, resembling fens in every respect, and it is by piercing the ground underneath their feet that they procure the fish hidden in these retreats.

Nevertheless, such floating shores can only exist on a small number of coasts, where the physical circumstances are quite exceptional; usually it is by grottoes and caverns hollowed out of the solid rock that the waters of the ocean penetrate far into the land. It is not to be doubted that there are below the level of the sea multitudes of those rocky galleries, but only those are known that are open to the strike of the waves, like the azure grotto of Capri. Lower down, the water closes the entrance to the lateral caverns, which will doubtless long remain unknown to us. But if we cannot explore grottoes still filled by the sea, we can at least see on elevated coasts like those of Scandinavia, immense caverns which the waves once freely traversed. One of the most imposing grottoes in the whole world is that which penetrates the splendid rock of Torghatten, rising like an enormous pyramid to more than 900 feet, on an island of northern Norway. This gallery, through which seamen see the light glimmering, is of an astonishing regularity. The thresholds of the immense portals, one of which has an arch of nearly 234 feet and the other of nearly 144 feet span, are found on each side to have the same elevation of 375 feet above the level of the sea. The ground, covered with fine sand, is almost level, and formed like the floor of a tunnel, where carriages might roll. The lateral walls present almost throughout a polished surface, as if they had been cut by the hand of man, and rise vertically to the spring of the arch; only towards the centre of the grotto the

* R. Thomassy, *Bulletin de la Société de Géographie*, 1864.

vault is less elevated than at the two extremities. Seen through this gigantic telescope, 900 feet long, the promontories, islets, innumerable reefs, and the thousand white crests of the breakers, form a spectacle of incomparable beauty, especially when the sun illumines the whole landscape with its rays.*

When the waves of the sea cannot enter into the caverns remote from the shore except by narrow channels, it often happens that a rivulet of salt water regularly flows towards the interior of the land, without ever returning to the ocean. This strange fact, which may seem at first sight a reversal of the laws of nature, may be observed on various points on the coast of calcareous countries, and especially on the coasts of Greece and the neighbouring islands.

Near Argostoli, a commercial town in the island of Cephalonia, four little torrents of sea-water, rolling on an average 55 gallons of water per second, penetrate into the fissures of the cliffs, flow rapidly among the blocks that are scattered over the rocky bed, and gradually disappear in the crevices of the soil. Two of these watercourses are sufficiently powerful to turn throughout the year the wheels of two mills constructed by an enterprising Englishman. Though the subterranean cavities of Argostoli are in constant communication with the sea, and the entrance to the canals is carefully freed from the seaweed that could obstruct the passage, or at least retard the current, the waters are not the same height in the grottoes as in the neighbouring gulf. This is because the calcareous rocks of Cephalonia, dried on the surface by the sea-breeze and the heat of the sun, are pierced and cracked throughout by innumerable crevices, which are so many flues aiding the circulation of the air, and the evaporization of the hidden moisture. We can compare the entire mass of the hills of Argostoli, with all their caverns, to an immense *Alcaraza*, the contents of which are gradually evaporated through the porous clay. In consequence of this constant loss of liquid, the level of the water is always lower in the caverns than in the sea, and to restore the equilibrium, the brooklets, which are fed by the waves, descend incessantly by all the fissures towards the subterranean reservoirs. It is probable that the constant evaporation of the salt water has resulted in the accumulation in the cavities of the island of enormous saline masses. Professor Ansted has calculated that the discharge of the two great marine streams of Argostoli would be sufficient to form each year a block of more than 1800 cubic yards of salt.†

* Vibe, *Küsten und Meer Norwegens, Mittheilungen von Petarmann*, 1860.
† *The Ionian Islands in the year* 1863.

CHAPTER XIX.

UNDERMINING OF ROCKS.—VARIED ASPECT OF CLIFFS.—PLATFORMS AT THEIR BASES.—RESISTANCE OF THE COASTS.—BREAKWATERS FORMED BY THE RUBBISH. —HELIGOLAND.—DESTRUCTION OF LOW SHORES.

All the rocky promontories exposed to the violence of storms, or simply washed by a current, are undermined at their base. The wearing away is accomplished in a more or less rapid manner, according to the progress of the waves, the distribution and inclination of the strata, the hardness of the rocks, and their chemical composition. The method of destruction depends at the same time on various hydrological and geological conditions. Strange as this assertion may appear, the water of the sea can even in certain cases destroy the rocks on its borders by combustion. Thus, the cliffs of Ballybunion, on the western coast of Ireland, long presented the appearance of a rampart of smoking lava. Those rocks which the waves of the Atlantic have pierced with grottoes, and sculptured in massive and fantastic forms, having one day fallen down very extensively, the alum and iron pyrites, which are contained in considerable proportion in the rocks, were exposed to the action of the atmosphere and the sea-water. A rapid oxydation took place, and produced a heat sufficiently intense to set the whole cliff on fire. For weeks the rocks were burning like a vast coal fire, and masses of vapour and smoke rose like clouds above the high wall besieged by the surf. Scattered around the space where the fire had prevailed, a heap of melted scoriæ, and clay transformed into brick by the violence of the fire, was to be seen.

Such is the diversity of destructive agents employed by nature, that, as we can easily understand, the aspect and form of the rocky coasts varies likewise in a remarkable manner. Thus the cliffs of England and Normandy, which are composed of somewhat friable rocks, fall when their lower strata are eaten away, and their sides being occasionally interrupted by "valleuses" (narrow openings where temporary or permanent brooks flow), they resemble enormous walls from 150 to 300 feet high. In the islands of the Baltic Sea, the chalky rocks, less exposed to the fury of the tempests than those of western Europe, are also less abrupt, and forests of beech-trees descend like sheets of verdure over the ruins of the cliffs. Elsewhere, especially on the coasts of Liguria, the promontories, formed of limestone rocks harder than chalk, do not fall in when their lower strata are carried away by the sea, and the waves, incessantly excavating the bases of these rocks, may carve them into colonnades, arched gateways, winding galleries, and vast grottoes, where the trembling water lights up the vaults with its azure hues. Other cliffs, of which the promontory of Socoa, near St Jean-de-Luz, may be considered as a type, are composed of slate rocks, variously inclined towards the sea. Worn away by the waves, some of the layers of schist are detached, others bend and part from each other, like the pages of an open book, allowing the water to glide in long foaming sheets into the very heart of the cliff, to spring up again from it in immense spouts. Finally, on other coasts the rocks cut by vertical fractures are gradually isolated from one another, and separated into distinct groups by the action of the waters. Surrounded by a roaring sea, they rise on their rocky bases like towers, monstrous obelisks, gigantic arcades, or crumbling bridges. Such are the innumerable rocks which tower above the waves in the archipelago of the Orkney and Shetland Islands. Black, slender, and enveloped with spray as with smoke, these wrecks of ancient cliffs justify the name of "chimney-rocks" which the English have given to many of them. On the northern coasts of Norway, not far from the polar circle, a rock rises in the midst of the waves, more than 900 feet high, which resembles a giant cavalier; hence its name of Hestmanden.

We see that the rocks which the sea-wave has eaten away are very various in form. Still we may say, as a general rule, that the inequalities of cliffs are in direct proportion to the hardness of the strata. The grooves that the waves slowly hollow out in the surface of the rock, the cavities that they scoop out in it, the arcades and grottoes which they excavate there, are the deeper the harder the stone is, for the beds of less solid formation fall in as soon as the lower layers are eroded. That part of the cliff which is only wetted by the foam and the mist of minute drops is less cut up than the base, and the grooves are less numerous there; but no vegetation as yet appears. Higher up a few lichens give a tint of greenish gray to the stone. Finally, those bushes which delight in breathing the salt air of the sea make their appearance in the angles and on the cornices of the rocks. It is at 100 or 120 feet high that this vege-

FORMATION OF CLIFFS WITH PLATFORMS.

tation begins to show itself on the cliffs at the border of the Mediterranean.*

Notwithstanding the astonishing variety of aspect presented by cliffs composed of various substances, chalk, marble, granite, or porphyry, we can still observe one trait of singular resemblance in the form of the rocks, which are covered by the waters of the sea at the foot of the abrupt walls. This feature consists in the existence of one or two platforms of varying dimensions, situated at the base of the escarpments. On the coasts of the Mediterranean and other seas with a very slight tide, where the level of the waters hardly varies excepting under the influence of the winds and storms,

Fig. 61.—Cliff on the Mediterranean.

there is but a single one of these platforms, while on the shores of the ocean, where the tides attain a height of at least several yards, two steps, one above the other, extend below the wall of the cliffs. When the rock is very hard the platforms present but a few yards

Fig. 62.—Ocean Cliff.

---

* Boblaye et Virlet. G. Collegno; *Geologia dell' Italia*.

in width, and perhaps may then be compared to a narrow cornice, suspended at mid-way between two abrupt walls, that of the cliff, and that which plunges into the abyss of the water. On the other hand, when the rocks are easily cut, the terrace of one or several stages over which the waves roll, has sometimes many hundreds of yards in width. At Inishmore, on the western coast of Ireland, the cliff presents a succession of regular steps like those of a staircase cut out for giants. The highest step, all encumbered with blocks, is that attained by the waves during a tempest; lower down are those bathed by the spring-tides, and then that where the ordinary tides are arrested. Still lower are the intermediary terraces, and the last two steps of the staircase are those where the water breaks during ordinary ebb, and at the low tides of the equinoxes.*

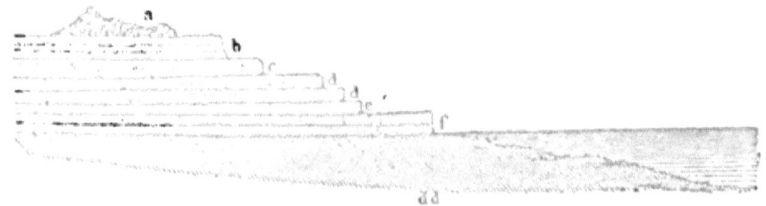

Fig. 63.—Tides of Inishmore. (Kinahan.)
a. Deposit of tempests.   *b.* Terrace of equinoctial tides.   *c.* Terrace of ordinary high tides.
dd. Intermediary terraces.   *e.* Terrace of ordinary low tides.   *f.* Low equinoctial tides.

It will be easily understood that these submarine ledges were formerly embedded in the thickness of the rock; they have resisted the assault of the waves, while the higher strata, sapped at their base more or less slowly, have fallen into the water. As the force of the waves is felt much less in the mass of waters than on the surface of the sea, the rock only allows itself to be cut into at the place where the breakers dash. But its submerged slopes remain relatively intact, and maintain more or less exactly the ancient outline of the coast. This is the reason why there exists on the shores of the Atlantic, and other seas, the level of which oscillates alternately with the ebb and flow, two platforms, one above the other, which correspond, the one with the level of low water, the other with that of high water. At the time of flow, the waves, urged by the tides, and more often too by the wind which accompanies the tide,† dash impetuously against the rocky walls, and push on vigorously their labours of destruc-

* Kinahan, *Geological Magazine*, August, 1866.
† See below, the section entitled, *The Air and the Winds.*

tion. During the period of the ebb, on the contrary, the water which breaks on the shore is retained by the current of low water, and is as though attracted towards the open sea: neither does it attack the cliff with as much energy as the rising tide. The difference which exists between the force of the waves of the flow and those of the ebb, can be measured by the respective extent of the intermediary platforms.

If the waves march constantly to the assault of the shore to transform into cliffs the heights of the coast, the latter, on their side, are not satisfied with merely resisting by their mass, and by the greater or less hardness of their strata, but many of them besides take care, one might say, to protect their threatened base against the waves. A thick vegetation of seaweed, like floating hair, drapes the cornices, breaks the force of the surf, and changes into torrents of eddying foam the enormous rollers which rush to attack the rocks with great speed. Besides, all that portion of the rocks comprised between the levels of high and low water is covered with balani and other shells, numerous enough to give the stone the appearance at certain hours of a swarming mass, and to form it afterwards into an immense immovable carapace.*

The coasts thus protected are precisely those which, by the solidity of their rocks, would best resist the attacks of the sea. As to the cliffs composed throughout their thickness, or only at their base, of less resisting materials, they give way too often for the molluscs and seaweed to venture in great numbers on that part of the rock which the waves have just assailed. Great blocks detach themselves from the upper strata, and fall on the beach. Afterwards, under the action of the waves, they break into smaller pieces, then into pebbles which the surge rolls and chafes incessantly. Under these fragments, constantly moved by the wave, no germ of animal or plant can develop itself, no living organism brought from the open sea can exist there. A desert is made even in the waters which dash against the roaring mass.

When this is the case, it is the crumbling masses and the pebbles of the strand which themselves serve as bulwarks of defence to protect the wall of the cliffs from fresh damage. Supported in a slope on the lower part of the rock, or else scattered in the waves and transformed into shelves, the fallen blocks break the force of the waves, and retard the progress of erosion. It is thus that on the coasts of the Mediterranean, near Vintimillia, the lower strata of the cliffs are

* See below, the sections entitled, *The Earth and its Flora*, and *the Earth and its Fauna*.

composed of a sandy clay, which the rain alone suffices to wash away, and this gives rise to a talus of masses of solid conglomerate detached from the upper layers, which thus protects the cliffs from the fury of the waves. In the same way, on the sterile shores of Brittany, the blocks of granite, cracked in all directions, and converted into shingle which the sea carries away and returns again, maintain intact during centuries the walls of rocks of which they formerly made a part.

The cliffs of Normandy, composed of materials much less hard than those of the promontories of Brittany, are also more easily worn away; still we must attribute their rapid erosion principally to the coastal current which carries away the shingle accumulated at the base of the rocks. The talus of fallen blocks constitutes at first a perfectly sufficient defence against the fury of the waves: but little by little, the chalky part of the rock is dissolved and deposited here and there on the mud-banks, while the masses of flint disengaged from the substance of the stone, cease to present a sufficient resistance to the waves, and are carried away into the neighbouring bays in immense processions parallel to the shore. On the south coast of England, the current of the coast is much less energetic, and the talus can in consequence long resist the attacks of the sea. A few years ago the waters undermined with a threatening rapidity the base of the cliff which rises not far from Dover, on the western side, and which the English have consecrated to Shakspeare, in remembrance of the beautiful description which he has given of it in *King Lear*. To preserve this historical promontory, the houses that it supports, and the railroad which runs through it in a tunnel, they formed the plan of blowing down the upper part. In the presence of an immense crowd, assembled to see this new spectacle, they fired hundreds of pounds of powder buried in a mine, and enormous masses of rock fell with a crash from the top of the hill; and now the force of the waves is broken on their talus. Mr Beete Jukes thinks that during 18 centuries this cliff and the neighbouring rocks have been worn away by nearly 1 mile.*

In the North Sea there is an island which by a singular misapprehension was believed to have been consecrated to Freya, the goddess of love and liberty, and whose ancient name of Halligland (land with the inundated banks) has been transformed for foreigners into that of Heligoland (holy land). The island, composed entirely of mottled stone, formerly surrounded by cretaceous beds, presents to the sea all round a cliff about 200 feet high, worn away at the base

* *School Manual of Geology*, p. 89.

ARTIFICIAL BREAKWATERS.

by the waves. By employing the heroic means which the English engineers have applied to the defence of Shakspeare's Cliff, and which the garrison of Heligoland had also inaugurated in the year 1808, by bombarding a crumbling cliff,* the inhabitants might surround their island with a great circular breakwater. But this dike would not last long, for the strata of mottled stone do not contain those beds of pebbles which serve to form shingle for a beach.

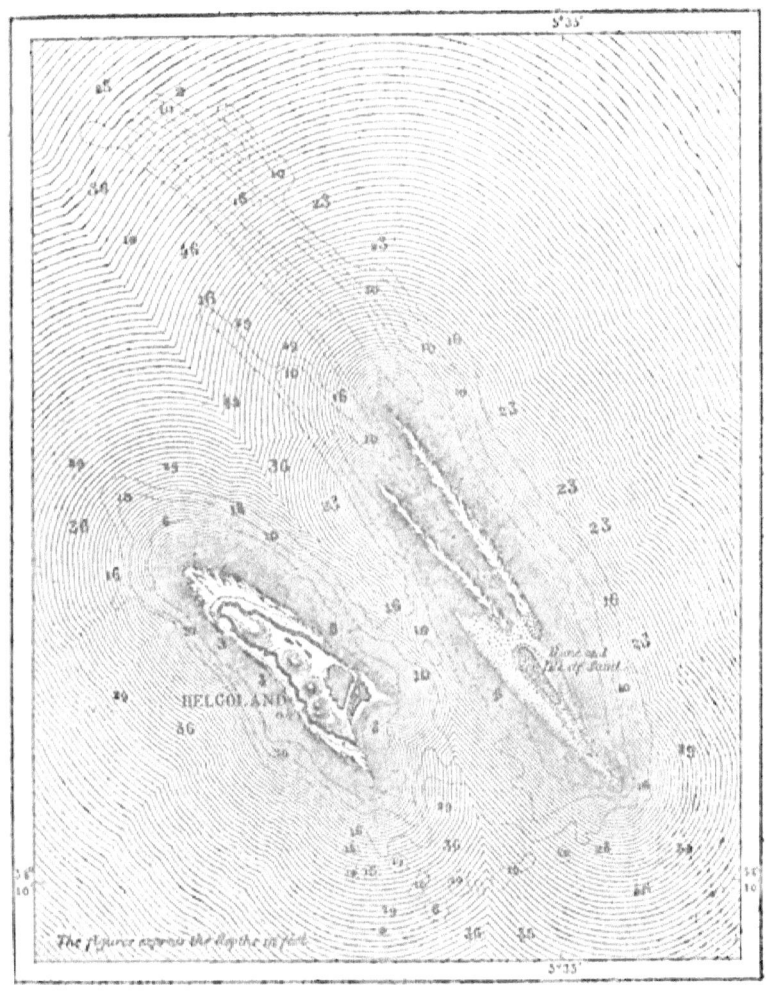

Fig. 64.—Heligoland.

---

* Hallier, *Nord-See Studien*, p. 73.

All the blocks would soon be dissolved by the waves, and not a single fragment remaining to protect the lower strata of the cliff against the destructive action of the waves, the work of erosion would freely resume its course. Devoted to certain destruction, the island is gradually melting in the waters, like an immense crystal of salt.

The learned do not all give the same degree of confidence to the documents relative to the ancient extent of Heligoland. Some, such as Wiebel,* regard those testimonies of the past as if destitute of sufficient authenticity, and think that the lessening of the island is accomplished very slowly. Others, on the contrary,† more respectful to the affirmations of the chroniclers, believe that in the space of five centuries the island has diminished by at least three quarters. However it may be, it is certain that the partially inundated lands, to which the island owes its name, have long since ceased to exist. It is equally certain that towards the end of the seventeenth century, an isthmus still united Heligoland to another islet, the cliffs of which rose to about 100 feet in height, like the principal island: two excellent ports, which gave the island a great strategical importance, opened to the north and south between the two rocky masses and their submarine extensions. The eastern island has now disappeared, and its cliffs are replaced by a few dunes and sandbanks, uncovered at low water: the ports no longer exist, and the ships of war drawing most water, can sail freely where the isthmus still existed less than a century and a half ago. Who would now recognize in this rock of Heligoland, hardly 1½ mile long, and about 2000 feet broad, the land of which Adam de Brêmse speaks in 1072, and which was then "very fertile, rich in corals, in animals, and birds," and which extended, says Karl Müller, "over a space of 900 square kilometres." ‡ In the present day, a few rows of potatoes and a few meagre pastures are the only remains that testify to the ancient fertility of Heligoland.

If the sea thus destroys countries bordered all round with rocky promontories, it respects still less the low strands, which in consequence of some modifications in the geography of the coasts, or in the relief of the submarine banks, are situated across the currents. In the very front of Heligoland, the shores of Hanover, Friesland, and Holland, which formerly seemed to sink gradually,§ offer the most striking example of this destructive power of the sea.

* *Die Insel Helgoland.* † Von Manck *Zeitschrift für die algemeine Erdkunde*, 1860.
‡ *Die Gefahren der schleswigschen West kuste:* Natur, March, 1867.
§ See in Vol. I. the section entitled, *The Slow Oscillations of the Terrestrial Soil.*

INUNDATIONS OF THE COASTS.

During sixteen hundred years, that is to say, ever since written history commenced in these countries, the life of the inhabitants of the shores has been nothing but an incessant strife against the encroachment of the waters. During this period the great irruptions of the sea may be counted by hundreds, and among these there are some which, according to the chronicles, must have drowned whole populations of fifty and a hundred thousand souls. During the course of the third century, tradition tells us that the island of Walcheren was separated from the continent; in 860 the Rhine rose, inundating the country; the palace of Caligula (*arx britannica*) remaining in the midst of the waves. Towards the middle of the twelfth century the sea made a new irruption, and the Lake Flevo was changed into a gulf, which was still more enlarged in 1225, forming the Zuyder Zee, that vast labyrinth of sandbanks, which, from a geological point of view, is still a dependency of the continent, and is separated by a long row of islands and dunes from the domain of the ocean. In the first years of the thirteenth century the gulf of Jahde was opened at the expense of the land, and never ceased to enlarge itself during two hundred years. In 1230 the terrible inundation of Friesland took place, which is said to have cost the life of a hundred thousand men. The following year the lakes of Haarlem overflowed the ground, then gradually increasing, united with each other to expand into an inland sea towards the middle of the seventeenth century. In 1277 the gulf of the Dollart, which is nearly 22 miles long, and 7 miles wide, began to be hollowed out at the expense of the fertile and populous countries, and transformed Friesland into a peninsula. It was only in 1537 that they could arrest the invasions of the sea, which had devoured the town of Torum and 50 villages. Ten years after the first invasion of the waters in the Dollart, an overflowing of the Zuyder Zee drowned 80,000 persons, and changed the configuration of the Dutch coast-line. In 1421, 72 villages were submerged at once, and the sea on retiring left only an archipelago of marshy islands and islets, covered with reeds and banks of mud, in the place of fields and groups of habitations: this is the country known under the name of Biesbosch (forest of reeds). Since this epoch many other hardly less terrible catastrophes have taken place on the coasts of Holland, Friesland, Schleswig, and Jutland.*

Of the row of 23 islands, which extended along the shore 15 centuries ago, only 16 fragments remain, and many are nothing

* Von Hoff; von Maack; Beyer; Baudissin; Karl Müller, &c.

else than simple ridges of sand. The Island of Borkum, as is shown by maps with less than a century's interval, has been singularly less-

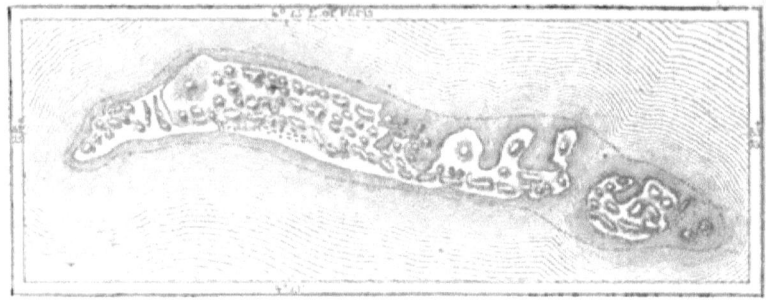

Fig. 65.—Isle of Borkum in 1738.

ened; the Island of Wangerooge, the wreck of the antique country of Wangerland, which was once united to the continent and extended

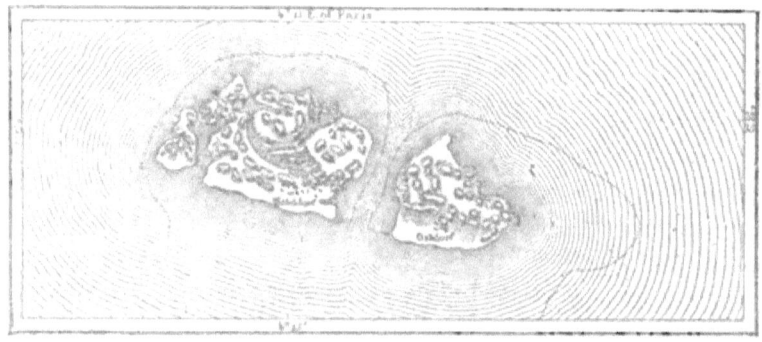

Fig. 66.—Isle of Borkum in 1825.

far into the sea, was in 1840 still a flourishing and populous island, and during the summer the bathers visited it in crowds. Now it is a strand of mud almost entirely abandoned. The Island of Nordstrand has diminished by eleven-twelfths since the commencement of the seventeenth century, and of the 24 islets which surrounded it 300 years ago, there only remain 11: the lead, when thrown out at the spot which was formerly in the centre of the island, indicates a depth of 7 fathoms. The Island of Sylt, and other lands of the coast of Schleswig, have been also much worn away, and it is known that in 1825 the sea opened a way quite across the peninsula of Jutland by hollowing out the Strait of Lymfjord.*

* Müller *Gefahren der schleswigschen Westküste: Natur*, March, 1867.

CHAPTER XX.

NORMAL FORM OF SHORES.—CURVES OF "GREATEST STABILITY."—FORMATION OF NEW SHORES.—COAST RIDGES AND SAND-BANKS.—INLAND BAYS.

THE shores most violently attacked by the sea are, generally, those which present the most indentations and promontories. The waves break most of all against the advanced capes that jut out farthest into the domain of the waters; but in proportion as the points retreat before the tide which wears them away, the destructive power of the waves diminishes; its force will even in time be reduced to nothing, when the base of the cliffs is sufficiently eroded, and describes no more than a slight curve in front of the coast. In fact, the outline of the coast which offers the greatest resistance to the assaults of the sea is not a straight line as one might suppose, but a series of regular and rhythmical curves.* The waves do not cease to labour at remodelling the shore as long as the latter does not present a succession of creeks gently curved from promontory to promontory. Each one of these rounded bays reproduces in large the form of the wave which breaks on it, drawing on the sand of the beach a long elliptical curve of foam-flakes.

The coasts of mountainous countries, to which the sea has already given the desired contours, unite an extreme grace with an admirable majesty. Such are the coasts of Provence, of Liguria, of Greece, of the greater part of the Iberian and Italian peninsulas. There, every promontory, the remains of an ancient chain of hills razed by the waves, lifts its terminal point in a high cliff; each valley which descends towards the sea terminates in a beach of fine sand, with a perfectly regular curve. Abrupt rocks and gently sloping beaches alternate thus in a harmonious manner, while the various geological formations, the greater or less width of the valleys, the towns scattered on the heights, or on the low shores, the curvatures of the coast, and the incessantly changing aspects of the water, introduce variety into the whole landscape.

* Elie de Beaumont, Baude, &c.

Sandy shores, as well as rocky coasts, have a normal profile composed of a series of bays and points. But these points, the relief of which is modified by every wave, are generally more rounded in their extremity than the promontories of rock. The monotonous coast of the Landes, which extends over a length of nearly 140 miles, from the mouth of the Gironde to that of the Adour, may be taken as a type of the shores which the waves of the sea model at their will. On these coasts the uniformity of the landscape is complete. The traveller in vain hastens forward; he might believe he had hardly changed his place, so immutable does the aspect of the scenery remain;

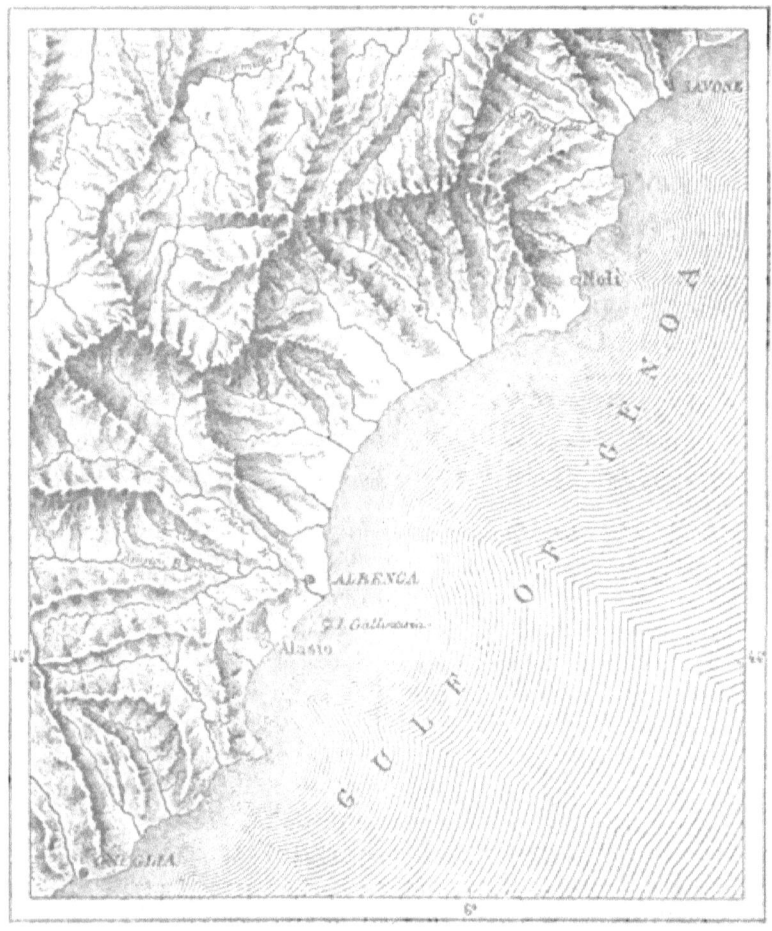

Fig. 67.—Curves of the coast between Oneglia and Savone.

always the same dunes, the same shells scattered on the sand, the same birds assembled by thousands on the edge of the lagunes, the same lines of waves which pursue one another, and break with great noise in a sheet of foam. In the whole field of view the only remarkable objects are the spars of shipwrecked vessels that can be seen from afar on the white sand. However, the shores which present the most regular series of convex and concave curves, which one might call the outline of the greatest stability, are exposed also to rapid erosions when the bulwark of defence which flanks them at either of their extremities yields to the pressure of the waves. Thus the shores of Medoc, which are the continuations of the uniform coast of Saintonge, to the south of the bay of the Gironde, have incessantly retreated before the sea ever since the rocky promontory (of which the ledge of Cordouan is the sole remains) disappeared under the united ravages of the river and the ocean.

But if the sea demolishes on one side, it builds up on the other, and the destruction of the ancient shores is compensated for by the creation of new coasts. The clays and limestones torn from the promontories, the shingle of every kind which is alternately thrown up on the shore and swept back in the waves, the heaps of shells, the silicious and calcareous sands formed by the disintegration of all these fragments, are the materials employed by the sea for the construction of its embankment, and the silting up of its gulfs.

It is on each side of the cliffs or low points worn away by the waves, that the work of reparation commences. Each wave accomplishes a double work, for in sapping the base of the promontory it loads itself with fragments which it deposits immediately on the neighbouring strand; by the same action it causes the point to retreat, and the shore of the bay to gain. Thus, owing to two series of apparently contrary results, the razing of the points and the filling up of the bays, coasts more or less deeply indented, gradually acquire the normal form with gracefully rounded curves. Whatever be the outline of the primitive coast, each inflection of the new shore rounds itself like the arc of a circle from promontory to promontory. In those places where the ancient coast was itself semi-circular, the sand or gravel cast up by the billows is deposited on the beach; but when the coasts are irregular and indented by deep creeks, the sea simply leaves them and constructs sand or shingle banks in front of them, which end by becoming the true shore.

The formation of such a breakwater may be explained in a very simple manner. The waves of the open sea, driven against the shore, first strike the two capes placed as guardians at the two extremities of the bay; here they break their force, and are thrown back against the tranquil waters of the bay. Thus arrested in their speed, they deposit the earthy matters which they hold in suspension, and also the heavier fragments torn from the neighbouring promontories. At the entrance to the fjords of Scandinavia, of Terra del Fuego, and all the other mountainous countries with deeply indented shores, the clear and deep water of the open sea only brings with it a relatively small quantity of *débris*, and can only form a submarine bank * from point to point. But along the lower coasts where the tide drives before it masses of sand and clay, the ramparts of alluvium constructed by the waves emerge gradually from the bosom of the waters.

Under the alternate influence of the ebb and flow, the sand and shingle are gradually deposited against the rocks of the capes, and thus they form at the entrance of the bay true jetties, the free extremities of which advance to meet each other. Being elongated unceasingly, the two segments end by uniting midway between the two capes, and thus form a large arc of a circle. the convexity of which is turned towards the ancient shore. The most furious assaults of the sea only serve to consolidate these banks by bringing other materials to them, and raising them above the level of the tides.

All these sea-banks present a geometrical regularity in their section; their form is, so to say, the visible expression of the laws which govern the undulation of the waves. Most often that part which fronts the sea is composed of several separate slopes which correspond to the different levels of low water, high tide, and storms: but all these beaches present uniformly a graceful curve, modelled by the breakers. At the base of the embankment the slope is very slight, and simply continues the declivity from the bottom of the sea: but it rises suddenly at an angle that is sometimes not less than from 30 to 35 degrees. Immediately beyond this edge a counter-slope begins where the upper curve of every high wave spreads in a foaming sheet. Further on rises a second talus, which the stormy waves sometimes strike and consolidate. The inclination of this second stage which looks towards the sea is very slight. From this side, the materials of the embankment, sheltered from the force of the wind, and from the violence of the waves, are gradually heaped up, and may even at length be covered by a bed of vegetable earth. Above this level dunes rise, or else we find the surface of the ancient bay transformed

* Darwin, *South America*, p. 24.

into a lagune. This outline of the shores is represented by the accompanying illustration, where the heights are strongly exaggerated.

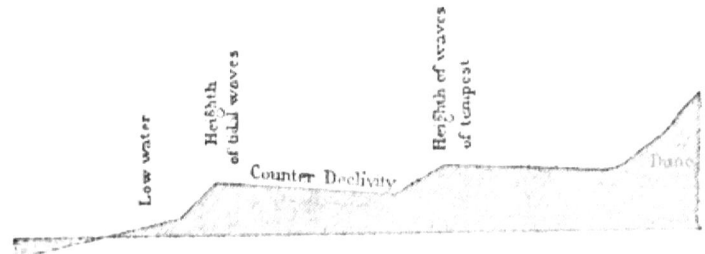

Fig. 68.—Section of sea-shore.

Notwithstanding the looseness of the materials which compose them, the banks are more solid than the promontories of rocks against which they rest, and when the cliffs have been levelled by the waves, the banks of sand again extend from one ledge to the other. They can be displaced by the influence of the currents and the winds, but they do not the less continue apparently immovable and more durable than the mountains. They do not, however, present a continuous development. When the inland bay is fed by one or several rivers, the mass of water which is discharged in this closed basin must necessarily break a passage to the sea, and pierce this ridge at the spot where it offers least resistance; that is to say, most often at one of its extremities. A remarkable example of this phenomenon is to be seen in Corsica at the mouth of the Liamone. In countries where the year comprises a dry period and a rainy season, most of the lagunes on the coast are alternately completely separated from the sea, and united with it by temporary embouchures of inconsiderable depth. When the mass of rain waters has flowed away, the breaches in the broken bank are instantly restored by the waves. In the same manner on the shores of seas with strong tides, a number of rivers are alternately canals of almost stagnant water, which a bank of sand separates from the ocean, and vast estuaries up which a powerful tide from the open sea flows. Thus, the Bidassoa, separated from the gulf at low water by a most graceful curved sand-bank, is at the hour of high water an arm of the sea, from 2 to 3 miles wide. Almost all the small water-courses which discharge themselves into the Atlantic are alternately rivers and marshes twice a-day. Even the Orne itself, whose large delta spreads like a fan beyond the coast-line, is lost in a shingle-bank at the hour of low water.

168   THE OCEAN.

If the permanent or periodical water-courses open for themselves a passage through the bank, these very rivers, on the other hand, serve to bring the inland shore and the sea shore gradually closer

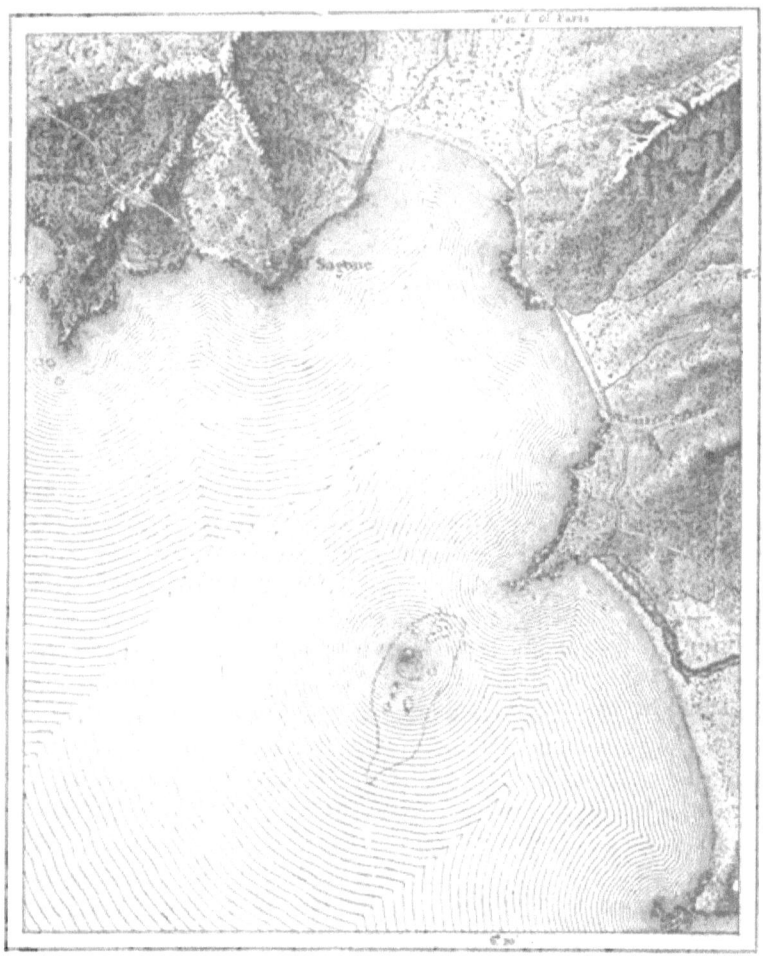

Fig. 69.—Mouth of the Liamone.

together by depositing their alluvium in the interior lagunes. The reeds and other plants which delight in the turbid waters contribute also to the transformation of the ancient bays into marshes and firm ground. Beds of vegetable detritus accumulated in the bays during a succession of years and centuries, rise in time above the ordinary level of the waters; and then come great trees which solidify the soil, and attach it definitively to the continent. In the

Fig. 70.—Mouth of the Bidassoa.

tropical regions, it is the different species of baobab and mangroves that aid most in the formation of the new shores. Raised on the scaffolding of their high roots, like aerial buttresses one above another, they grow in the midst of the lagune. Hidden by the floating forest, the muddy liquid is soon filled with rubbish; the branches and uprooted trunks of the trees, being much heavier than the water, incessantly raise the bottom, and end by appearing on the surface. A new vegetation immediately takes possession of this yet undecided shore.

The same hydrological laws which determine the formation of banks between two capes are at work to bring about the same result between two islands, or an island and the mainland. On the coasts of

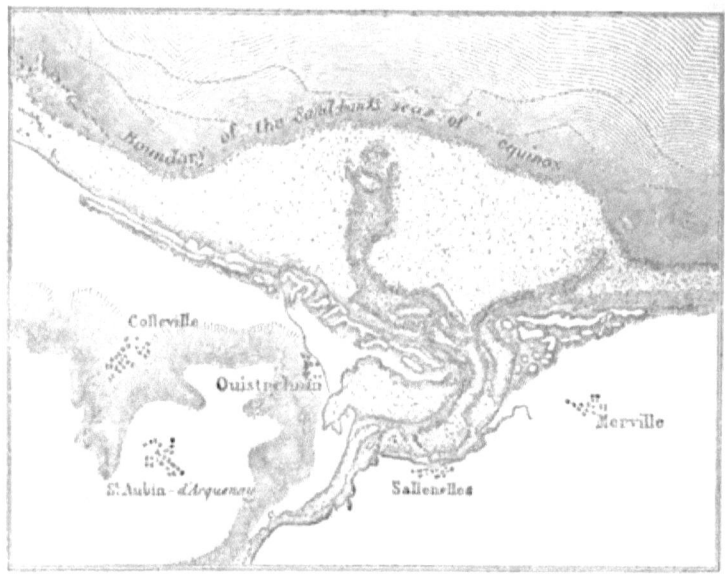

Fig. 71.—Mouth of the Orne.

Europe a great number of coast lands have thus lost their insular character, and are changed into peninsulas; the strait has been gradually changed into an isthmus. The peninsula of Giens, between Hyères and Toulon, presents a remarkable example of this transformation. It is connected with the continent by two banks of fine sand, above 3 miles long, each developed in regular curves, which turn their concave faces towards the open sea. Between these two banks stretches the vast lagune of Pesquiers. At the sight of this inland sheet of water and these low shores, hardly elevated above the level of the Mediterranean, one cannot doubt that the mountainous peninsula of Giens was formerly an island like Porquerolles or Port-Cros, and that the two roads, now separated, of Hyères and Giens, were formerly one strait. The two uniting banks which joined the ancient island to the coast of Provence, have been raised by the waves in the same manner and on the same plan as the coast ridges of the continent. As to the differences of appearance, they can all be explained by local circumstances. Thus the bank which the isthmus of Giens turns

FORMATION OF PENINSULAS.

towards the west is composed in reality of two unequal fragments, due to the existence of a submarine reef which breaks the force of

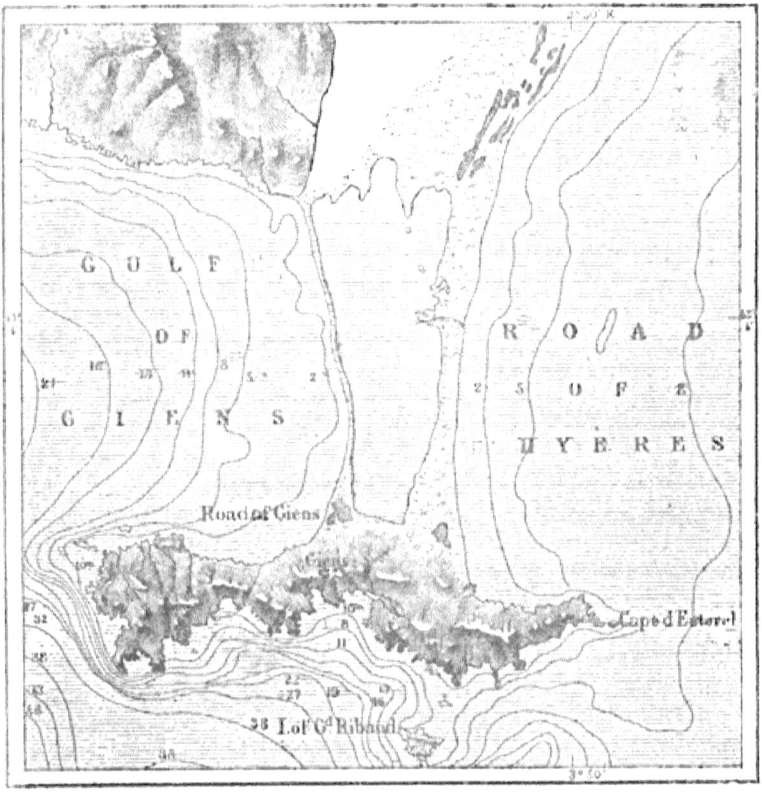

Fig. 72.—Peninsula of Giens.

the waves at a little distance from the strand. It is equally to local causes that we must attribute the inequality of thickness presented by the two ridges of the isthmus. Undoubtedly the eastern bank owes

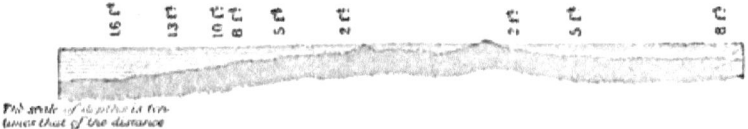

Fig. 73.—Section across the peninsula of Giens.

its greater solidity and height to the double action of the marine current that tends from east to west, and of the mistral which blows the opposite way in the direction from north-west to south-east; the two contrary forces have left as witness to their strife this rampart of sand and *débris*.

The peninsulas of Cape Sepet, near Toulon, of Quiberon, in Brittany, of Monte Argentaro, on the coasts of the Tyrrhenian Sea, and others less known, have been united to the continent by similar connecting causeways analogous to those of Giens. There, too, the two armies of waves which break in the midst of the strait have gradually erected between them a double wall of separation, consisting of banks of sand and shingle. There, too, the two semicircular jetties have drawn nearer together in their central convexity, and the two triangular spaces, which separated the respective extremities, have at first been occupied by lagunes. In our days, most of the ponds, gradually filled up by sand, have been transformed into marshes or covered by dunes; the two littoral ridges have been mingled in a single one. Thus, the narrow isthmus of Chesil-Bank, which extends over a

The Depths are represented in fathoms.

Fig. 74.—Peninsula of Cape Sepet.

length of 16 miles, between the coast of England and the former island of Portland, is composed of a single bank of shingle. In the same manner, the two French islands of Miquelon, near Newfoundland, which were still separate from each other in 1783, have been united since 1829 by a rampart of sand, which the waves of two opposite gulfs have erected conjointly.* Guadeloupe is likewise an example of this phenomenon of junction between two lands of distinct origin. The range of volcanic mountains which rise to the west is united to the low island of the east, and the two islands are now joined to each other by a marshy plain, where the waters of the small canal, called the Salt River, stagnate. In the two islands of Choa-Canzouni, bathed by the waters of the Indian Ocean, an analo-

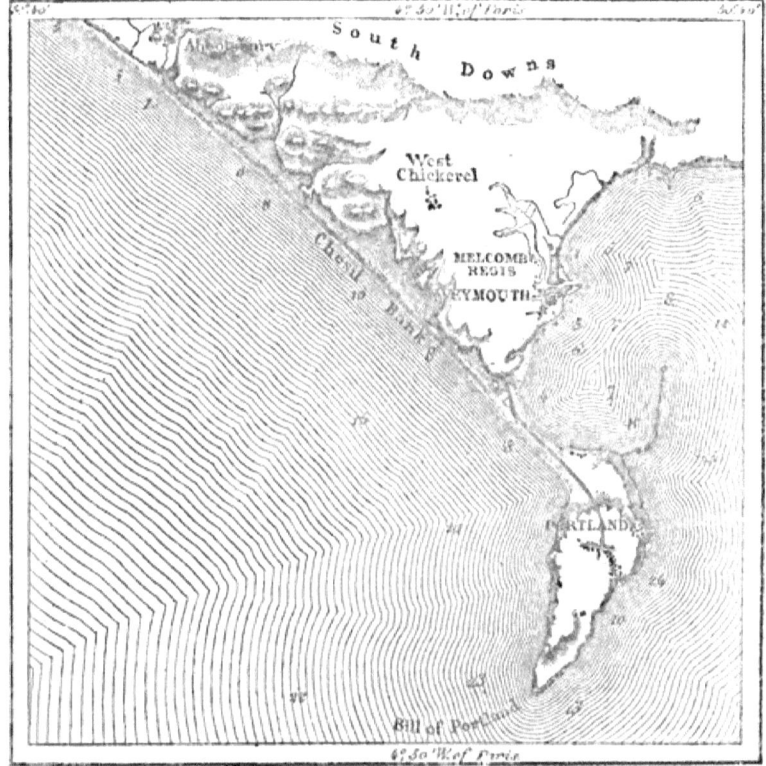

Fig. 75.—Chesil Bank.

gous phenomenon is presented, but there the connecting bank between the two islands is reduced, so to speak, to a mathematical point.

M. Elie de Beaumont estimates the length of the coasts which owe their present configuration to banks of shingle and sand to about one-third of the total development of the continental shores. It is in

* Brué, *Bulletin de la Société de Géographie*, 1829.

these basins with slight tides that these ridges present the most con-

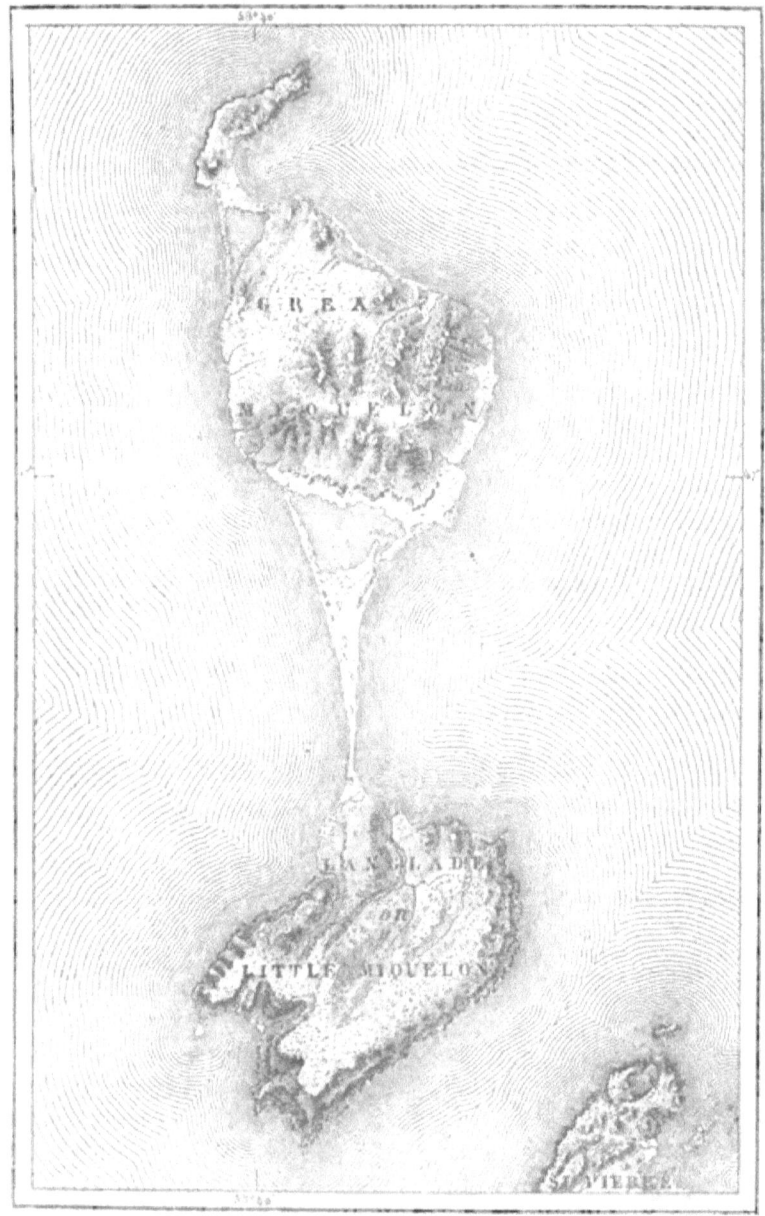

Fig. 76.—Miquelon Isles.

siderable dimensions. In France, all the shores of the Gulf of Lyons, from Argelez-sur-Mer to the mouths of the Rhone, form a series of

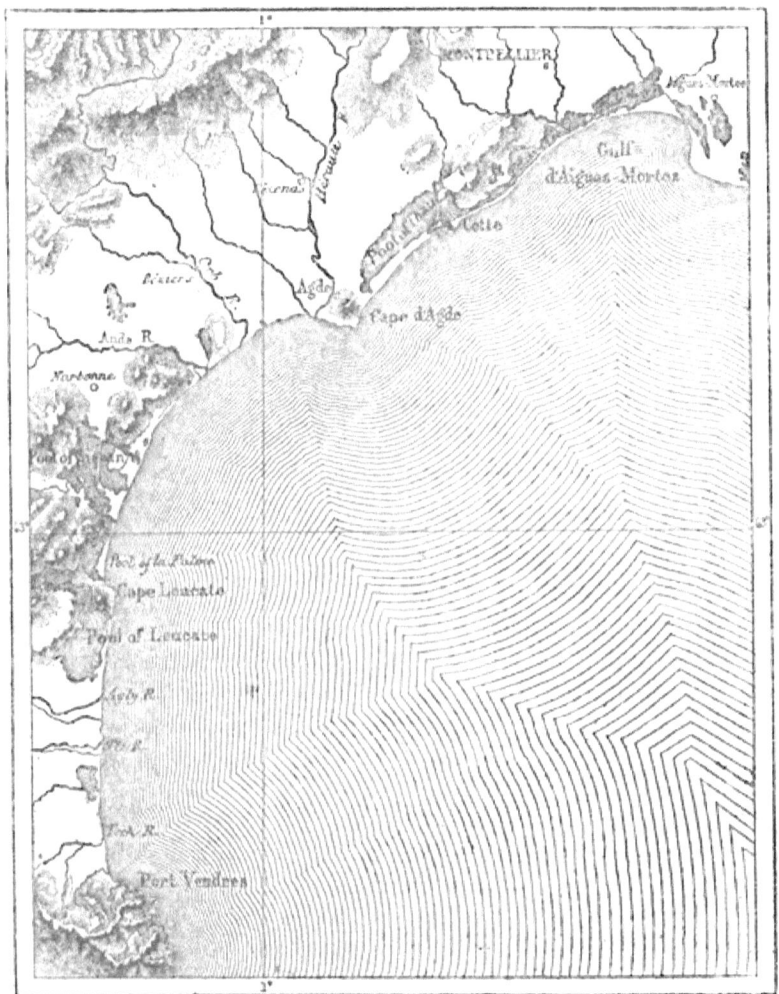

Fig. 77.—Coast-ridges between Port-Vendres and Aigues-Mortes.

coast-ridges, only interrupted by the rocks of Leucate, of La Clape, Agde, and Cette, and developed in a vast semi-circle nearly 125 miles long. The numerous ponds or *étangs* which it now separates from the Mediterranean, and which the alluvium of the rivers, the marine sand, and invading agriculture, are unceasingly transforming into solid earth, were doubtless so many bays along the base of the hills of Languedoc. Even since the historical epoch, these inland waters have notably diminished in extent, and vast gulfs, changed into marshes, to the great detriment of the public health, have poisoned the atmosphere with their miasma. That which contributed most actively to the diminution of the surface of the pools were the *graus*, or passages by which the water of the sea brought in heaps of sand during tempests. These openings, some temporary, and others permanent, but enlarging and diminishing by turns, and changing place, now in one direction and now in another, do not cease to modify the condition of the *étangs* and countries on the coast. Here they allow masses of water to break through, which submerge the shores and excavate the soil, and elsewhere they are obstructed, and spread banks of fetid mud as far as the eye can see before the villages of the coast. In order to prevent for the future the transformation of the *étangs* into mud and marshes, M. Régy has proposed to replace the old tortuous *graus* by channels for drainage, which, during the fine weather, allow the lacustrine waters and those of the sea to communicate freely, but the sluices of which would be closed during storms.

The *lidi* of Comacchio, as well as those of Venice, and of the ancient city of Aquileia, restrict the basin of the Adriatic, which formerly penetrated much further into the lands to the west and north-east. On the southern coasts of Brazil and the Guinea coast, the littoral ridges thus cut off considerable tracts from the ocean; but nowhere are these *levées* of sand seen more numerously and better developed than around the Gulf of Mexico, and on the eastern coasts of the United States. We may say that over a length of about 2500 miles the outline of the American continent is formed of a double coast, the one bathed by the sea and the other by the interior lagunes. In front of the ancient coast, with its irregular indentations, the new shore describes its graceful curves from promontory to promontory, and not even allowing itself to be arrested by the mouths of the rivers, stretches across the outlets in dangerous bars.

Thus the indented coasts of North Carolina, and the ramified gulfs which cut into these peninsulas, and are prolonged even into the interior of the land, in the form of marshes, are masked on the side next the sea by a natural bank nearly 220 miles long, against which the most fearful waves of the northern Atlantic break. These banks so gracefully curved are not constructed by the sea alone. They are due also to the pressure of the fresh waters brought from the Alleghanys by the Neuse, the Tar, the Roanoke, and other

HEADLANDS OF NORTH CAROLINA.

rivers;[*] the direction of the breakwaters indicates precisely the line of equilibrium between the marine and fluvial waters. Within the

Fig. 78.—Lagunes and Lidi of Venice.

outer coast-line it has been possible, without any considerable artificial means, to put the whole series of interior lagunes in communication

[*] See in Vol. I. the section entitled *Rivers*.

with one another, and thus to allow ships to make long sea-voyages completely sheltered from storms. Even the *marigots* of Guinea, which spread parallel to the shore, have at all times served to facilitate traffic between the peoples of the coast; but it is said that these marshy canals are gradually being filled up, either by the activity of the vegetation, or because of the sand which the wind of the desert transports thither.*

Much less extensive than the banks of the Gulf of Mexico and of Carolina, those of the eastern Baltic are not less curious by the geometrical regularity of their forms, and, besides, they have been the object of long and serious study. Three great rivers, the Oder, the Vistula, and the Niemen, discharge themselves each into a vast lagune or *Haff* (*hafen*, port), which a narrow tongue of land, called there a *Nehrung*, separates from the open sea. The *haff* of the Oder,

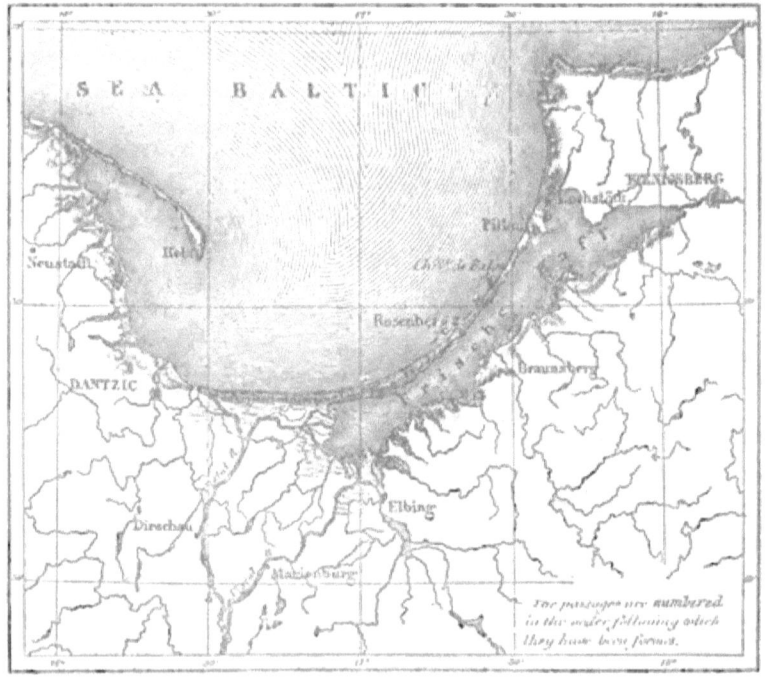

Fig. 79.—Coasts of Dantzig and Pillau.

the entrance to which is guarded by the town of Swinemünde, is already in great part filled up by mud. The *Curiche Haff*, or

* Borghero, *Bulletin de la Société Géographie*, July, 1866.

haff of Courlande, is much freer from alluvium, and the Nehrung which defends it is a narrow sand-bank about 68 miles in length. The central *haff*, known under the name of the *Frische Haff*, is protected by a bank similar to that of Courlande, but still more regular. All the western part of the estuary has already been filled up by the alluvium of the Vistula, the waters of which have opened a way through the bank. This embouchure has often changed its place. Till the fourteenth century it was to the north of the present passage, near Lochstädt (town of the gap or *grau*). Later, it opened at Rosenberg, nearly in the middle of the dike. To preserve their commercial monopoly, the merchants of Dantzig filled this opening up by sinking five ships there; but another passage was formed almost immediately at a little distance towards the north, near the castle of Balga. More greedy than wise, the Dantzigers again attempted to arrest the waters of the Vistula, and closed the passage of Balga. It was then that the *Nehrung* was broken before Pillau. Since this period the passage has not been sensibly displaced, and Pillau has always remained the port of the *Frische Haff*. To the north of Dantzig a curious bank, 20 miles long, unites the mainland to the picturesque island of Hela (the holy). Doubtless the ancient inhabitants of the country experienced a sentiment of religious terror at the sight of the rude waves which assail this wooded hill, united to the continent by this narrow dike of sand stretching far away into the dim distance.

It is to the same order of phenomena that we must refer the gradual prolongation of tongues of land, which, bathed on either side by a current, project to a great distance into the open sea, owing to the fresh materials which each new tide adds to the terminal point. It is thus that in less than 16 years Cape Ferret has advanced about 3 miles across the channel by which the basin of Arcachon communicates with the open sea. In 1768 the Cape was almost to the west of the basin properly so called. In the latter part of the eighteenth century, and at the commencement of the present, the winds from the north, which blow in those parts more frequently than the other atmospheric currents, had caused the dunes of the promontory to advance each year in a southerly direction, while the surf from the open sea, and the ebb of the basin, incessantly added fresh masses of sand to the point. In 58 years, from 1768 to 1826, the Cape lengthened by above three miles towards the south-east, with an average speed of 94 yards per year, or about 8 to 10 inches per day. The point increased, so to say, visibly; but a few years later the passage had sud-

180   THE OCEAN.

denly changed its direction, and tending to the north, the tidal current commenced to wear away the peninsula, and gradually caused it

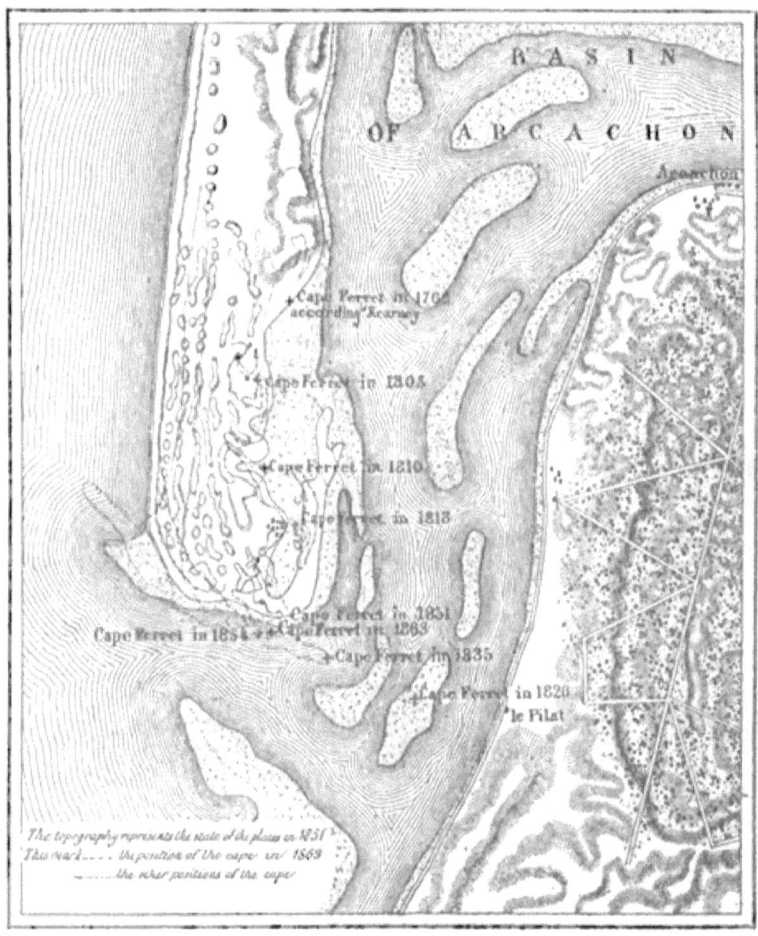

Fig. 80.—Different positions of Cape Ferret from 1768 to 1863.

to retreat towards the north-west. In 1854 the extremity of the cape had retrograded nearly 1¼ mile. It is said that it is at present nearly stationary, but if the channel moves towards the south, which may happen any day, it is not to be doubted that the point of the cape would recommence encroaching upon the sea in the same direction.

CHAPTER XXI.

SHALLOWS OF THE COAST.—DEPOSIT FROM CALCAREOUS ROCKS.—APPEARANCE OF STRANDS AND BEACHES.

THE formation of shallows and sand-banks is connected also with that of littoral ridges; they are developed parallel to the shore, under the combined influence of the currents along the coast and winds from the open sea. A glance at a chart which indicates the form of those ramparts hidden under the waves, shows at once that all these invisible banks of sand and mud tend to elongate

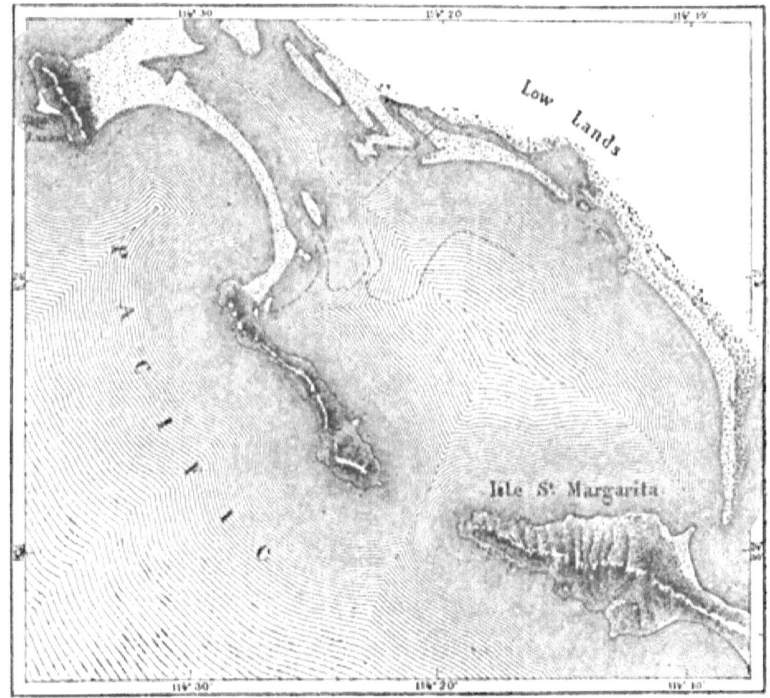

Fig. 81.— Road of La Madeleine, California.

themselves in a straight line, or to follow graceful curves no less regular than those of the littoral ridges. In all the gulfs and straits on the coasts of California, the Carolinas, and Brazil, in the Channel and in the North Sea, there exist along the coasts a multitude of these banks, the arrangement of which indicates exactly the path of the contrary or parallel currents which have been formed by their meeting. Their depth varies. There are some over which large ships can sail without danger; but there are others very near to the surface of the water, over which the waves incessantly break. It is these banks, hardly below the level of the sea, which are the most dreaded; and the English and American sailors, thinking of the fate that perhaps awaits them on these hidden sands, have gaily given them the ironical name of "frying pans." In wide-

Fig. 82.—Gulf of Carentan.

mouthed gulfs, and along straight coasts, the sea endeavours to construct new shores by means of deposits of mud. The remains of seaweed and animalculæ, mixed with sand and clay, are deposited in deep layers on the coast, and cause the outline of the shores to advance gradually. Mud has been accumulated by hundreds and tens of thousands of millions of cubic yards since the historical era in the ancient Gulf of Poitou, in the Gulf of Carentan, situated at the foot of the peninsula of the Cotentin, in the bays of the Marquenterre and of Flanders, in certain estuaries of the Netherlands and of Friesland. In these parts the sea and the land are mingled; the sea, gray or yellowish, resembles an immense slough, and continues the oozy surface of the shores; one does not know where the water terminates, or where the plain of mud, incessantly stirred by the tidal wave, begins. Still the mud which emerges at low water is, little by little, heaped up and consolidated; a species of conferva covers its surface with a slight carpeting of a pink hue; then come the herbaceous salicornia, which contribute to elevate the soil by their stiff branches springing from the stem at right angles. To this first vegetation succeed other marine plants—carices, plantains, reeds, and climbing trefoil. Then is the time to recover the oozy meadow for agriculture, and to connect it with the mainland by defending it with strong dikes against the assaults of the sea.*

In the seas whose waters have a high average temperature, the waves do not confine themselves to constructing littoral ridges and filling up the bays; they even build actual ramparts of stone. In consequence of the rapid evaporation produced by the rays of the sun, the calcareous particles and mud contained in the water are gradually deposited along the shores and over the base of the promontories. Mixed with sand and fragments of shells, it tends to form solid shores with regular contours. On the Atlantic coasts of France, at Royan for example, one can, here and there, already observe some formations of this kind; and further to the north, at Elsinore, some of these stones have been discovered, containing ancient Danish coins.† On the French shores of the Mediterranean these modern rocks are very numerous, and in a short walk one can often collect a large quantity of sandy blocks and various conglomerates, united by calcareous substances, and containing multitudes of broken shells. The Museum of Montpellier possesses a cannon which was discovered near the principal mouth of the Rhone, imbedded in a calcareous deposit. On the

* Emile de Laveleye, *Revue des Deux Mondes*, Nov. 1, 1863. Von Manck, *Zeitschrift für die allgemeine Erdkunde*, Jan., 1860.
† Von Hoff, *Veränderungen der Erdoberfläche*, tome iii. p. 311.

northern coasts of Sicily, where the mean temperature of the waters rises to 64·4 Fahr., the stones and pebbles of the shore are, in many places, agglutinated by calcareous cement.* In the same way the fragments of rocks, which the torrents of Arabia Petræa bring every winter from the top of the mountains to the shores of the Red Sea, are, in the space of a few weeks, converted into a stratum of solid conglomerate. Every year a new layer of stone is added to the old ones, and in future centuries we shall be able perhaps to estimate the age of the formation by the number of its beds, one above the other, in the same way as we recognize the age of a tree by the number of its annual rings of wood.†

We must explore the shores of the Antilles, or other tropical seas, to observe this phenomenon of the formation of rocks in all its grandeur. There the waves, heated to 89·6 Fahr. by the rays of a vertical sun, deposit limestone in sufficient quantity notably to increase the extent of the shore. The tufa of Guadeloupe, in which the famous human skeleton exhibited at the British Museum was found, belongs to this recent formation. It grows, so to say, under the very eyes of the observer, and gradually covers with a rocky crust all those objects which the sea rejects, and which the brooks bring down from the interior. In many parts of Terra Firma these quarries of marine stone are actively worked for building towns on the coast, and all the excavations made in these banks of limestone are soon filled up by new materials. The quarry grows under the labourers who are occupied in detaching the blocks; hence the name of *Maçonne-bon-dieu*, which the natives have given to those rocks which seem to be renewed of themselves.

On the shores of Ascension Island Mr. Darwin found some of these conglomerates cemented by marine limestone, whose specific weight was 2·63, that is to say, hardly less than that of Carrara marble. These beds of compact stone, deposited by the sea, contain a certain quantity of sulphate of lime, as well as the animal substances which are evidently the colouring principle of the whole mass. Sometimes the translucid stucco covering the rocks has the polish, hardness, and hue of nacre; moreover, as chemical analysis proves, this kind of enamel and the envelope of living molluscs are composed of the same substances equally modified by the presence of organic matter. Mr. Darwin has seen some of these calcareous deposits,

---

* De Quatrefages, *Souvenirs d'un naturaliste*.
† Marsh, *Man and Nature*, p. 455.

DRYING UP OF BAYS AND SALT LAKES. 185

whose composition and nacreous appearance seem as if they ought to be attributed to guano saturated with salt water.*

This construction of new shores, either by the sea itself or by the coral animals, like the gradual formation of the dunes, results in completely modifying the form of the coast, by separating from the rest of the sea large bays, which the rapid evaporation transforms later into firm land. It is thus that on the eastern coast of Africa the small Lake of Bahr-el-Assal, at the extremity of the Gulf of Tedjura, has been separated from the sea by a slender ridge of sand, and dried up under the rays of the sun. Rain-water being very rare in this country, and the basin receiving no affluent, its waters have not

Fig. 83.—Bahr-el-Assal and the Gulf of Tedjura.

been replaced, and now it is only a marshy hollow, the level of

* Darwin, *Volcanic Islands*, p. 49, and following.

which is about 570 feet below the Red Sea.* While they occupied the coast of Abyssinia during the last war, the English engineers discovered another basin, dried up and completely covered with salt, which was 189 feet below the sea-level. It is very probable, too, that the depressions in which the great river Haouach loses itself to the south of the plateau of Habesch, are likewise below the sea-level. The Isthmus of Suez formerly offered a phenomenon similar to that of the bank of Tedjura. There, too, a lacustrine sheet, which previously formed part of the sea, had been enclosed by the littoral ridges, and had almost entirely evaporated. Only in our days the grand inter-oceanic canal causes the marine waters to flow again through this dried-up lake. The ancient banks on the shores of the Mediterranean and the Red Sea, which the forces at work in the interior of the planet had gradually elevated to the height of several yards,† have been pierced by engineers, and an artificial strait much more important for human progress than the great arm of the sea was formerly, joins the Mediterranean with the Arabian Gulf.

If the great geological labours of the ocean, such as the erosion of cliffs, the demolition of promontories, and the construction of new shores, astonish the mind of man by their grandeur, on the other hand, the thousand details of the strands and beaches charm by their infinite grace and marvellous variety. All those innumerable phenomena of the grain of sand and drop of water are produced by the same causes which determine the great changes of the shore. At the sight of the delicate lines which the dying wave traces on the beach, as well as in the presence of the wild coasts which the surf wears away in fury, we feel ourselves brought back by various impressions to the contemplation of the same general laws. Each wave accomplishes on its little portion of the shore a work similar to that of the great ocean on the outline of all the continents. In a space of only a few yards we can see the regular curves of small bays round themselves, littoral ridges rise, inland lagunes form, and cliffs of shells and fuci being eroded. At the extremity of certain sheltered bays, in the Bay of Beaulieu, near Nice, for example, black masses of from 3 to 4 yards in height may be seen, cut into peaks and pierced with caverns like rocks; these are masses of algæ.

Among the various marvels of the shore nothing astonishes us

* Rochet d'Héricourt, *Voyage au Choa*. Christophe, *Journal of the Geographical Society*, vol. xii.

† See in Vol. I. the section entitled, *The Slow Oscillations of the Terrestrial Soil*.

more at first than the designs traced on the sand often with perfect regularity. Every breaker brings with it shells, pebbles, and other fragments of all kinds and of different sizes. These objects are so many little reefs which divide the wave on its return to the sea, and cause it to trace upon the ground a network of intersecting lines. The surface of the strand presents in consequence an interlacing of innumerable lozenges, all ornamented with a shell or pebble at their upper end, and pointed or slightly rounded. All these little lozenges are themselves comprised within large quadrilaterals formed by furrows having, as starting-point, an object of relatively considerable dimensions. Contrasts of colours aid in the relief to vary still more this varied aspect of the shore. The differently coloured materials being in general of a different specific weight, are distributed in a regular manner in the various parts of the lozenges. One side of the figure may be formed of small crystals of mica, while another is composed of black sand charged with peat, another of pink or yellowish shells, and the fourth of grains of a pure white. Sometimes the sand, impregnated with organic substances, shines like watered-silk, or is slightly iridescent, as if a very thin layer of oil were spread over the ground.

All these hues modify the aspect of the shores infinitely, and the greater or less inclination of the ground introduces yet a new element of variety into the network of lines. In all those places where the slope is considerable, the water hollows out the sands in the figure of miniature rivers with their tributaries and deltas. Besides, these small hydrographic systems themselves differ from one another according to the inclination of the ground and the weight of the grains of sand. In one place the sloping ground and the fineness of the displaced materials permit drops and streamlets of water to descend in a straight line towards the sea; in another the rivulets, making their way with difficulty between the obstacles that arrest them, flow in winding courses. Elsewhere, watercourses cannot even be formed. The water of the sea remains on a horizontal strand, and all the wavelets reproduce, in hollows or in relief on the sand of the bottom, all the movements which the breath of air impresses on them. There is no appreciable difference between the varied surface of the shore exposed freely to the wind, and that of the sand which a thin watery bed covers, excepting perhaps that the furrows of the pool are more regular and deeply hollowed out.

Among the innumerable phenomena that might keep a geologist all his lifetime on the sea-shore, we must include a kind of miniature volcano. In breaking regularly on the shore, the wave brings each time a certain quantity of sand which it spreads in a thin layer. The air absorbed into the pores of the soil, immediately disengages itself in bursting bubbles; but here there are always a great number of aerial particles which cannot penetrate the damp bed of sand, and remain enclosed. Under the influence of the warmth of the ground or of the surrounding air, these particles dilate little by little, the pressure of the gas raises the hardened coating, and forms a cone, which sometimes bursts in consequence of the inward pressure, and throws out in volleys little spouts of sand-grains. It is true that unobservant persons walk over millions of these humble volcanoes without perceiving even one, but those who love the earth under all its aspects, and who contemplate with the same admiration the grain of sand and the mountains, can easily discover and study them. For the naturalist, who sees immense forests in every heap of algæ and a world of animals in the remains which strew the sand, the thousand wonders of the shore cannot fail always to awaken an intense pleasure.

CHAPTER XXII.

ORIGIN OF ISLANDS.—ISLANDS OF CONTINENTAL ORIGIN.—ROCKS OF THE SHORES.—
ISLANDS OF DEPRESSION, ELEVATION, AND EROSION.—ISLANDS OF OCEANIC
ORIGIN.—ATOLLS AND VOLCANOS.

ON viewing the great geological labours accomplished by the dash of the waves on the various coast-lines, savants have often asked what is the share the sea takes in the formation of islands. Among the lands which are scattered over the surface of the ocean, some disposed in groups or series, and others completely solitary, how should we distinguish between those which the sea has detached from the continents, and those which have existed in an isolated manner from all times like separate worlds ? Is it even possible in the present state of science to attempt a classification of islands, according to their origin ? Yes, this work may be commenced. By calling to our aid the new resources which botany and zoology offer to physical geography, we may affirm that, sooner or later, we can indicate with certainty the manner of formation, and the relative age of each oceanic country.*

Firstly, it is evident that the islands, islets, and rocky ledges situated in the immediate neighbourhood of the coasts, are a natural dependancy of the continents, and geologically make a part of them. At the base of the high mountains, which send far into the sea advanced capes, similar to the roots of an oak, we may see in many places, so to say, the crests of the lateral chains continue under the surface of the ocean. The outline of the continental heights sinks by degrees ; to the mountains succeed the hills, and the promontory of rocks, whose escarpments plunge beneath the even surface of the waters. An inconsiderable strait, simply a hollow where the waves meet each other, separates the cape from a less elevated island. But further on there opens a wide channel, and the peak which shows itself at the surface on the other side of the submarine valley is no longer anything more than a needle of rock. Beyond stretches the open

* See especially the works of Darwin, Wallace, and a study by M. Oscar Peschel, *der Ursprung der Inseln; Ausland*, Jan. and Feb., 1867.

sea where the submerged ledges, if any still exist, are only revealed by the whitening foam. On all the abrupt coasts these islets belonging to the primitive architecture of the continent are very numerous, and even in certain parts form real archipelagos. Norway, Western Scotland, Chilian Patagonia, and all those countries where the fjords change the coast-line into an immense labyrinth, are thus bordered with innumerable islands, having likewise their indentations, their straits, and their girdles of islets. This is because, since the relatively recent retreat of the glaciers which filled all the space comprised between the circles of snowy plateaux and the exterior promontories, the original relief has but slightly changed. The terrestrial alluvium brought down by the torrents has only filled up a small number of valleys; and the bases of the islands and capes, plunging deeply into the waters, have not served as support to marine alluvium similar to that which spreads along the low coasts. Isolated rocks, which the ice formerly surrounded as it now surrounds the "Jardin" of Mont Blanc, now rise in the midst of the waters, but they are not the less the salient points of the continental relief; in shallower waters, where the deposit of the marine alluvium would be easily accomplished, they would long since be joined to the shore.

Among the islands which may be considered as simple dependencies of the great neighbouring lands, we must also class not only those which the marine or fluvial alluvium has raised, simple emerged banks which are especially found along low coasts and near the mouths of rivers; but likewise the islands which are due either to the rising or gradual sinking of the ground. Thus, the chain of insular downs which defends the coast-line of Friesland and Holland against the assaults of the North Sea, from Wangerooge to the Texel, is most certainly the remains of the antique shore; and it is this rather than the half-submerged beaches of the Dollart and the Zuyder Zee which marks the true boundary between land and sea.* On the other hand, the coasts of the Scandinavian peninsula, which rise slowly above the waves, have been enriched with new islands during the course of the present geological epoch. In the maze of the Norwegian fjords, in the Lofoten Isles, in the Archipelago of the Quarken, hidden ledges have become visible rocks, then extensive islands where the algæ have been gradually replaced by a terrestrial flora. While the continent was encroaching upon the sea, the islets here and there have risen up and spread far over the waters like the leaves of some gigantic plant. The insular rocks rise

* See above, p. 161.

slowly from the depths of the ocean, elevated by the same force which raises the neighbouring continent. And is not a like phenomenon accomplished on the coasts of Scandinavia? Perhaps even the large island of Anticosti, which extends in the Gulf of St Lawrence over a length of more than 125 miles, is one of these slowly elevated lands, for, according to the testimony of Prof. Yule Hind, one does not find in the granitic valleys of its rocks either serpents or batrachians, as on the coasts of Labrador and Canada. If it is really thus we could hardly admit that Anticosti has ever been in communication with the continent of America, it must have emerged from the waters like the islets of the Scandinavian coast-line.

It has happened differently with regard to Great Britain and the greater part of the islands which fringe the outline of the continent. It is certain that England formerly made a part of Europe. This is proved by the perfect agreement between the shores on each side of the Straits of Dover;* it is also proved by the fauna and the flora of the British Islands, in which all the animals and all the wild flowers are colonists from the neighbouring world; not a single species belongs peculiarly as its own production to the soil of old Albion.† In the same manner, Ireland has been separated from Great Britain during the present geological period, and around the two principal islands, a number of secondary fragments, the Isle of Wight, Anglesea, and the Scilly Isles, have been similarly isolated in the midst of the waves.

A multitude of islands, situated, like England and Ireland, in the neighbourhood of continents, are also simply fragments which the waves, aided perhaps by the gradual sinking of the land, have detached from the shores of the mainland. The magnificent Archipelago of Sunda, the Moluccas, and the neighbouring islands of Australia, present the most remarkable example of this breaking into pieces of the continental masses. A channel, nearly 19 miles wide, and more than 109 fathoms deep, passes between the two large islands of Borneo and Celebes, and continuing in a southerly direction, separates the two volcanic countries of Bali and Lombok, very near to each other. This channel is the ancient strait which served as the common limit to Asia and the southern continent. To the west, Java, Borneo, Sumatra, the peninsula of Malacca, and Cambodia, rest on a submarine plateau, which lies hardly 33 fathoms below the surface of the waters: to the east, Sumbava, Flores, Timor, the

* See in Vol. I. the section entitled, *The First Ages*, and above, p. 14.
† See below, the section entitled, *The Earth and its Flora*.

Moluccas, New Guinea, and Australia, are likewise on a sort of pedestal, which sinks gradually, and upon which the zoophytes construct here and there long barrier reefs. Thus, as the naturalist Wallace has demonstrated by his researches in the Indian Archipelago, all the species of plants and animals differ completely on each side of the dividing channel. The fauna and flora are Asiatic to the west, while to the east they present the Australian type; even the birds, for whom a strait a few leagues in width would seem but a slight obstacle, are distinctly different in each of the two groups of islands.

We must therefore see in the Australian Archipelagos the wreck of a great continental mass, which must have divided into numerous fragments at epochs more or less distant from our time. We may say as much of the islands of the Ægean Sea, of those of Denmark, of the Polar Archipelago, of the New World, of the maze of the Magellanic Islands, and of the greater part of lands which surround the shallower waters in the neighbourhood of the coasts. As to the great islands of the Mediterranean, Cyprus, Crete, Sicily, Sardinia, Corsica, and the Balearic Islands, they are also very probably the remains of more extensive countries formerly united to those continents now known as Europe, Asia, and Africa. For though these lands, with the exception of Sicily, all rise from the depth of abysses, having, on an average, from 500 to 1000 fathoms depth, nevertheless the fossil and living species of the Mediterranean Islands do not differ from those of the neighbouring continents, and it is consequently there that we must seek their origin. From a geological point of view, one can even say that the countries of the western basin of the Mediterranean, Spain, Provence, the Italian peninsula, Tunis, Algeria, and Morocco, form with the neighbouring islands a whole much more precisely defined than, for example, central Europe, from the Straits of Gibraltar to the shores of the Caspian. In spite of the depths which separated them, the coasts lying opposite to each other, on each side of the Tyrrhenian Sea, have preserved a similarity of physiognomy in the flora and fauna of the land.

The Mediterranean Islands may thus be considered either as dependencies of the neighbouring continents, or, better still, as the remains of an ancient country partially swallowed up. Still, there exist in the midst of the sea insular masses in which geologists see nothing else than the witnesses of continental tracts which now have disappeared. Thus Madagascar, though sufficiently near to Africa, seems a sort of separate world, having a flora and fauna belonging peculiarly to itself, and even possessing entire families, especially of serpents and lemurs, which have no other representatives in our planet. Strange to say, even the island of Ceylon, half united to Hindostan by the rocks, islets, and sand-banks of the Pont-de-Rama, differs much from the neighbouring peninsula by the general facies of its animals and plants, and we may question if, instead of being simply a dependence of Asia, it is not, on the contrary, the last remains of an ancient continent which extended over the area of the Indian Ocean, and comprised Madagascar, the Seychelles, and other islands now almost imperceptible on the map.

Among the fragments of vanished worlds, we ought also probably to class the greater part of the Antilles and New Zealand. The larger Antilles present a much more striking contrast with the countries of North America than that between Ceylon and the peninsula of the Ganges. By elevation and geological character, Hayti and Jamaica do not in any wise resemble the low lands of the American coast, situated on the other side of the gulf; their vegetable and animal species differ notably from those of the neighbouring continent, though winds, currents, birds of passage, and even man, have worked together for an unknown number of centuries to carry animals and plants from one shore to the other. As to New Zealand, it is quite a distinct world, whose flora and fauna have an essentially original character. Neither the fossil nor the living species resemble those of Australia or South America.* And the greater number of savants agree with the opinion of Hochstetter, who sees in New Zealand and in Norfolk Island the fragments of a continent isolated ever since the commencement of the Mesozoic period. While Great Britain may be considered as a type of the islands scarcely separated from the neighbouring continent, her fine colony at the Antipodes represents, on the contrary, an ancient world, gradually reduced by subsidence and the erosions of the sea, to the dimensions of a mere insular group.

The present shape of islands often allows us to recognize what was their earlier form when they extended over a much more considerable space. By their outline, and ramifications, the mountain-ridges indicate in a general manner the first configuration: they are as the fragments of a skeleton around which we reconstruct, in thought, the contours of the ancient continental body. Besides, many of these, of which only the primitive skeleton remains, and whose plains have disappeared, are indented in the most curious manner, and their shores often present the most fantastic outlines.

* See below the sections entitled, *The Earth and its Flora*, and, *The Earth and its Fauna*.

Thus, Choa-Canzouni, in the Archipelago of Comoro, is a group of two large islands, united by a sort of stalk; Nossi-Mitsiou, in the

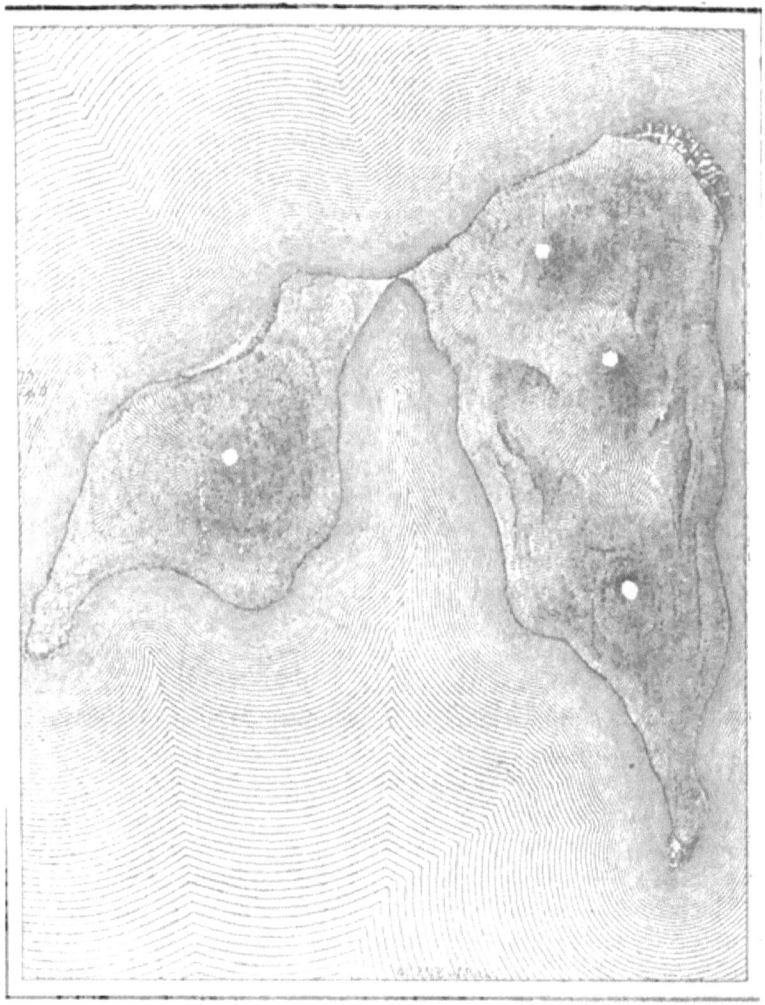

Fig. 84.—Choa-Canzouni.

same region, resembles a trunk with two broken boughs; finally, Celebes and Gilolo, so remarkable by the parallelism of their gulfs and promontories, seem to be both constructed on the same model; and what we know of the mountains of Borneo allows us to believe

FORMS OF INSULAR LANDS. 195

that if this large island were to be submerged beneath the sea, its

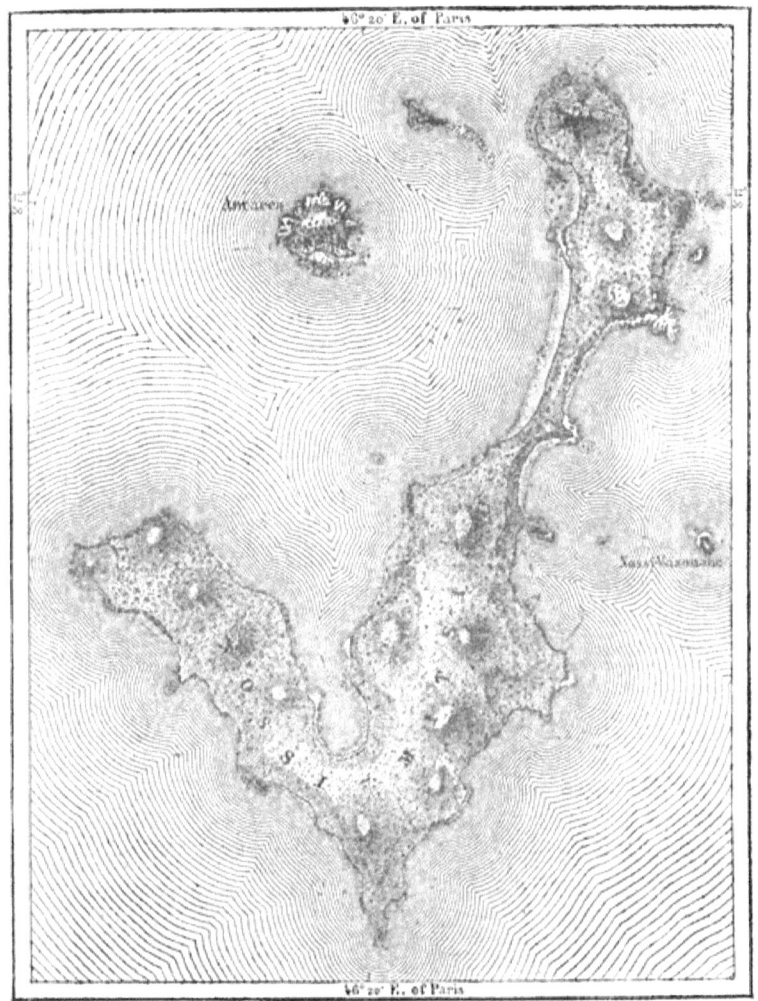

Fig. 85.—Nossi-Mitsiou.

shores would resemble, by their contour, those of its two neighbours in the sea of the Moluccas.*

Beside these fragments of ancient or modern continental masses, all the projections which show themselves above the level of the

* Oscar Peschel, *Ausland*, 1868.

ocean are either built by zoophytes, or else cast up by volcanos from the bottom of the sea; one or the other is, without exception, the origin

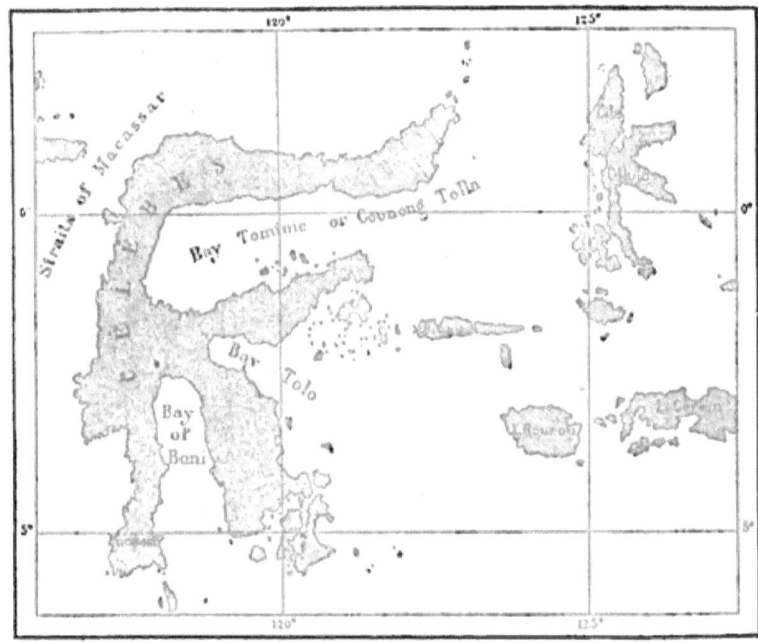

Fig. 86.—Celebes and Gilolo.

of all these islands. The first, as we know,* are disposed in *atolls*, or annular reefs, formed themselves of rings of smaller dimensions, while cones of lava, that are elevated in the open sea, rise proudly above the waves, and reveal the independence of their origin by a declivity which is continued pretty regularly below the waters. Still we can see by the example of the volcano of Stromboli, and more plainly

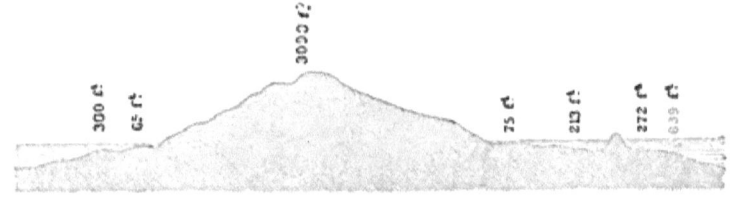

Fig. 87.—Section of Stromboli, from S.W. to N.E.

---

* See in Vol. I. the section entitled, *The Slow Oscillations of the Terrestrial Soil*; and below that entitled, *The Earth and its Fauna*.

ORIGIN OF ISLANDS.

still by that of the island of Panaria, that the waves constantly lessen the submarine slopes by distributing to a distance the lava and cinders rejected by the craters.

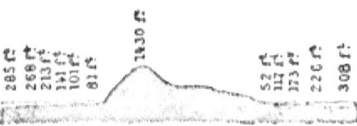

Fig. 88.—Section of Panaria, from N.W. to S.E.

Compared with the lands of continental origin, the truly insular masses composed of lava, or built by the coral animals, have relatively a very slight extent. It seems that, according to the general arrangement of the globe, the separation must at first have been much more marked between the sea and the emerged lands. On one side great continuous countries, on the other desert oceans, appears to have been the natural distribution. But the incessant work accomplished on our planet, as on all the stars of heaven, has infinitely modified the form of the continental surfaces and the channels which separate them. In the same way as by its rains and snows the sea has scattered lakes over the regions raised above its level, and traced the innumerable valleys of the water-courses, so have the lands given to ocean those myriads of islands and islets which vary its surface so gracefully. The alluvium of the rivers, the erosive power of the waves, the internal forces, which slowly raise or depress vast countries or cause cones of lava to spring up suddenly from the deep; finally, the numerous organisms which assimilate the various substances contained in sea-water, have all worked in concert to scatter here and there islands of different forms and sizes, some in larger, and others in smaller groups, or even completely isolated. Later, the winds, rains, monsoons, and other meteoric influences of the atmosphere; the oceanic currents, the ebb and flow, the undulation of the waves, all which moves and floats in the water and in the air, birds and fish, seaweed and drifted wood, foam and dust—have never ceased to act directly or indirectly, to introduce life into these islands, to people them with species of animals and plants, and thus to prepare them for the abode of man.

BOOK V.—THE DUNES.

CHAPTER XXIII.

DUNES RESULTING FROM THE DECOMPOSITION OF ROCKS.—FORMATION OF MOVING DUNES ON THE SEA-SHORE.—SYMMETRICAL DISPOSITION OF RIDGES OF SAND.

It is principally upon the sandy beaches of the ocean that those changing hillocks known under the name of *dunes* rise in long rows. Nevertheless, the phenomenon of the elevation of the sand in moving hills may also occur at a great distance from the present sea-shore. Dunes are formed on all points of the globe where the wind finds and drives before it light sandy materials; but we must remark that these substances only exist in considerable quantities on the shores of the sea and large lacustrine basins, at the bottom of ancient gulfs and straits transformed into deserts, on the banks of rivers, which roll sand along their beds, and which are exposed to frequent changes of level by the alternation of droughts and inundations. It is the waters which, by their destructive action on the cliffs, prepare the sandy particles necessary for the construction of dunes; and this origin allows us to consider the shifting ridges of sand, whatever be their distance from the shore, as products of the ocean.

In all the great deserts of Asia and Africa, we see some of these terrestrial waves caused by aerial currents.* Some exist also on the banks of the Nile and other great rivers. Even in France very fine dunes about 30 feet high rise on the banks of the Gardon immediately below the celebrated Roman bridge; it is the mistral which has raised them. In leaving the gorge which encloses it, this wind seizes the particles of fine sand left on the shores and dried by the sun, and deposits them at the entrance of the plain, where it spreads over a wider extent, and loses in intensity what it gains in surface.

A certain number of dunes have been formed on the spot during

* See in Vol. I. the section entitled *Plains*.

the course of centuries by the disintegration of freestone rocks.
Fogs, rains, frosts, and other atmospheric agents, gradually wear
away the stone and transform it into sand, which falling leaves
fresh beds at the surface. These are subject in their turn to the
destructive meteoric influences, and it is thus that the rock, once
solid, is gradually changed, often to a considerable depth, into a mass
of crumbling sand. The grains chafed against each other during their
fall become finer and finer, and when the wind is high, it can carry
away these sandy particles, cause them to ascend the slope of the talus,
and sometimes even raise them in clouds like the smoke of a volcano.
Nevertheless, the dune, still enveloping a solid kernel and composed
in great part of grains heavier than those of the sea-coast, is not
entirely displaced by the action of storms; it only takes another form
in consequence of the gradual change of its slopes. Several mountains of this kind near Ghadamès, which were formerly rocky hills,
rise to 150 and 600 feet high. One of them, which is not less than
510 feet, has an inclination of 37 degrees on the side exposed to
the wind; nearly the greatest slope that a talus of sand can present.*

As to dunes properly so called, those which are found far in the
interior of continents cannot be compared in importance with those
which are developed in long ridges, parallel to the sandy shores of
the sea. On the strands of the ocean which are not rocky the existence of dunes is almost constant; the only low shores which
are destitute of them are those which the waves have formed of
clayey substances, of compact mud, or sand much mixed with animal
and vegetable detritus. The sandy shores of the Mediterranean, of
the Baltic, and other inland seas, where the tides are hardly perceptible, also present very insignificant dunes, because the want of ebb and
flow does not allow the sand to acquire sufficient mobility. We see,
however, some more than 90 feet high between Vera-Cruz and
Tampico, on the shores of the Gulf of Mexico, where the tides are
very slight. On all oceanic coasts, the sand of which is loose enough
to allow itself to be raised by the wind, the formation of dunes is
accomplished with perfect regularity.

These hillocks rising, so to say, beneath the very eyes of the observer, it is not difficult to follow their progress, nor to offer a theory
regarding them. The waves constantly agitating the shifting foundation of the shore, become charged with arenaceous matters, and
spread them in thin layers over the strand. Then, at low tide,

* Vattone, *Mission de Ghadamès;* Barth, *Zeitschrift für die Erdkunde*, March, 1864.

the grains of sand soon become dry, and cease to adhere to each other, and thus allow themselves to be carried towards the land by the wind from the open sea. These are the materials of dunes. If the shore rises towards the interior of the continent in a perfectly even manner, this sand, cast up by the waves above the sea level, and carried far by successive gusts of wind, would extend over the ground in layers of uniform thickness; but the inequalities of the surface prevent this. Pebbles, branches and trunks of trees covered with shells, plants and bushes with tough roots, project above the beach, and oppose the advance of the wind, which glides over the ground, carrying the grains of sand that have remained on dry land. These slight obstacles suffice to determine the origin of dunes by obliging the breeze to let fall the little cloud of arenaceous or calcareous dust with which it is charged. The horizontality of the shore is thus interrupted; rows of sandy knolls, which are subsequently to rise to real hills, commence to be traced upon the ground.

When the wind from the open sea blows with sufficient force, we can not only witness the growth of the dunes, but we can also aid in their formation, and verify by direct experiment the assertions of theory. If we deposit some object on the ground, or, better still, thrust a row of stakes into the sand, perpendicularly to the direction

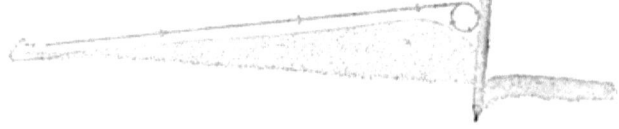

Fig. 89.— Formation of a Dune.

of the wind, the current of air which strikes against the obstacle will instantly rebound to form an eddy or whirlpool, the diameter of which is always proportioned to the height of the stake. Arrested by this eddy, the grains of sand carried by the wind are gradually deposited on the near side of the barrier, till the summit of the miniature dune is on a level with the imaginary line leading from the shore to the upper end of the obstacle. Then the sand driven by the breeze from the sea, which ascends the inclined plane presented by the front of the hillock, no longer allows itself to be carried in the eddy and brought back. It crosses the little ravine which the gyration of the air has produced in front of the palisade, and falls beyond it to accumulate gradually on the other side of the obstacle, taking the form of a descending talus (Fig. 90). It is due to the knowledge of

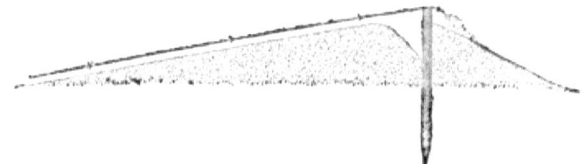

Fig. 90.—Formation of Sand Dunes.

these facts, that we are able to force the elements to construct a protecting rampart of dunes on various points of the coast threatened with erosion by the waves of the sea.*

Such are always the commencement of dunes, whatever be the object which opposes itself to the wind. It is easy to convince oneself of this by the sight of the houses or huts which the Custom-House officers and shepherds erect in the sandy hollows of the dunes of the Landes, not yet fixed by seedling trees. On the side towards the sea, which is also that from which the wind blows in terrible gales, the dwelling remains separated from the talus of sand by a ditch of defence, as regular as if it had been hollowed out by the hand of man; but on the side which fronts inland the sand is gradually heaped up, and if it is not swept away, does not fail to rise soon to the level of the roof.

On the slightly-undulating plateau which extends at the foot of the grand pyramids of Egypt we can also study the same phenomena. The winds from the east and north-east, which strike against the eastern face of the enormous masses of stone, rebounding and developing their reflected waves on the ground, do not allow the sand to be deposited on the lower steps of the edifices. It is only at a certain distance, at the precise spot where the current is neutralized by the masses of air coming directly from the east, that the dunes can form. To the west of the pyramids, on the other hand, long mounds of sand, more or less inclined, support themselves against the base of the monuments. In the same way, at the foot of certain cliffs of Liguria, where the sands accumulate in dunes, there always exists a sort of trench between the rock and the moving heap.

When the labour of man does not intervene to arrest the progress of the dunes formed on the sea-shore, the various obstacles which have determined the accumulation of the sands disappear at first on

* See below the section entitled, *The Work of Man*.

the descending side under successive beds; then when this part is entirely hidden, the front begins to be buried in its turn. The wind, instead of developing itself according to a horizontal plane, as on the surface of the ocean, is obliged to take an oblique direction to ascend the slope of the dune. As soon as it is sufficiently elevated, the atmospheric current passes freely above the obstacle which arrested it before, the little eddy which revolves in front ceases its gyrations, and nothing then hinders the sand from gradually filling up the ravine which the aerial current had maintained in front of the barrier. Soon the summit of the dune coincides with that of the obstacle: the latter disappears completely, and the hillock, growing like a wave which approaches the shore, and constantly raising its crest higher, which is incessantly displaced, continues to encroach upon the land. The various strata of sand which the wind from the open sea successively brings to the summits of the dunes, spread in large sheets over the descending talus, and glide down to the base. In the Landes of the Gironne the western slope of the dunes, whose base is not worn away by the sea, is, on an average, from 7 to 12 degrees. The eastern slope, which is that of the descending talus, is from 29 to 32 degrees; that is to say, three times as great. It would be 45 degrees if the rains did not make ravines in the talus and thus prolong the inclination.*

Thus the dunes incessantly gain, owing to the new layers of sand added to their changing talus. But the action of the prevailing wind does not limit itself to increasing them; it ends by displacing them entirely, and making them, so to say, travel over the ground. The object at the base of which the eddy of air had deposited the first grains of sand is at length decomposed; inclemencies of the weather, insects, moisture, and chemical agencies destroy it, and when it has disappeared the sand which it retained shifts again. The wind, which only carried away the superficial beds of the dune to replace them incessantly by new sheets of sand, can now carry away all the anterior part of the hillock; it lengthens the descending talus at the expense of the shore side, and the base of the hill, worn away by the wind, constantly retreats from the shore. The dune is on the march; it advances inland. Such is the mobility of the sands that even when the waves erode the foot of the dune, and force it to fall into the sea, the summit does not the less advance towards the continent. Destroyed on one side it invades on the other, like those voracious insects which, even when cut in half, do not cease to eat.

* Raulin, *Géographie Girondine*.

The high dunes of Lagrave, to the south of Archachon, are the most curious in this respect; below, the sea forces them to fall in; above, they bury the pine-trees in their invading masses of sand.

The most favourable days for observing the progressive march of dunes are those when a gentle breeze, strong enough however to drive the sand before it, blows in a perfectly uniform manner. From the top of the dune we see innumerable grains of dust swiftly scaling the slope. Glittering in the sun, and whirling like the midges in a fine summer's evening, they attain the summit, then accumulate in the form of a cornice on the other side of the ridge, and from time to time occur little falls, which spread over the surface of the talus, like sheets of water over the sides of a rock, and whose contours remind one of light draperies covering one another. When a high wind blows with violence, and in successive gusts, the encroachments of the dune are accomplished in a manner much more rapid, but often much more difficult to observe. The summits of the hillocks, which are enveloped in clouds of dust, resemble volcanoes vomiting smoke; the front of the dune is furrowed and scooped out by the wind; masses of sand, laden with marine remains brought by the storm, fall down with an audible sound, and are disposed in unequal layers over the descending talus. A section taken across a dune would permit us to count and measure the different strata varying in thickness and composition, which the winds have successively brought. Here we find a fine sandlike dust; there, a stronger wind was charged with a heavy shelly sand; while, again, a storm has carried away entire shells, branches, and waifs. However, the particles transported by the wind are, in general, all the finer the further they are from the sea, and this is reasonable, for they must

Fig. 91.—Section of a dune.

fly more easily the less resistance they offer to the aerial current which bears them. In the narrow rows of dunes which border certain parts of the coast of the Mediterranean, we can clearly see over a breadth of some hundreds of yards, the moving materials succeed each other, distributed according to their weight. First, there are the fragments of shells, then the large arenaceous débris, then the fine sands.*

If the inclined plane which the dune turns towards the sea remained perfectly even, the zone of the shore would only present, in all its extent, a single rampart of sand gradually encroaching on the lands. But at length the slope of each dune cannot fail to offer some inequalities caused by foreign bodies, or by plants that take their origin in the sand. All the salient points strong enough to resist the wind serve as supports to new dunes, grafted, so to say, on the sides of the ancient one. These new dunes themselves bristle with irregularities, which other sand-hillocks soon cover, and it is thus that all those ranges of moving hills arise, which are separated by long and narrow valleys, called *lettes* or *lèdes* by the peasants of the French Landes. In certain places, especially between Biscarosse and La Teste, these "lettes," for a length of several leagues, resemble the dried-up beds of large rivers, surrounding large islets of verdure with their sandy waves.

Notwithstanding the apparent disorder of these hillocks, in the midst of which an inexperienced traveller might easily lose his way, the general disposition of the sands can always be referred to a uniform type, which local geographical facts variously modify, such as the contours of the marine shore, the nature of the soil, the force and direction of the winds, the presence or absence of vegetation. The dune nearest to the sea, and, in consequence, the most recent, is less elevated than the older hillock situated immediately beyond; and this in the same way attains a less considerable height than the following hill. In a system of dunes, generally each range which is developed further inland exceeds the preceding ones in elevation, and forms, as it were, a new step on the slope of the great primitive dune which serves as an *avant garde* to the army of sands. This last dune, the true crest of the entire system, enlarges itself, little by little, with all the materials which have served for the formation of the inferior dunes situated on the side nearest to the sea. The grain of sand which the air carries to the summit of the first hillock, and which falls afterwards into a ravine, may remain during centuries under the superincumbent masses; but owing to the constant progress of the dune, the superficial layers of which are swept by the wind and then let fall by it further down the talus, this grain of sand at last reappears, is carried anew to a summit, it descends again, and thus does not cease to travel from dune to dune, to the last.

* Marcel de Serres, *Bulletin de la Société Géologique de France*, 1859.

As the innumerable arenaceous particles are moved by virtue of rigorous laws, we can consequently measure the force of the winds, by the height of the mass, and the rapidity of the displacement of the hillocks. Attentive observation permits us, in the same way, to compare with each other the various atmospheric currents which drive the sands onward, and to indicate exactly the one whose action is the most energetic. Thus, in the Peninsula of Arvert or La Tremblade, situated between the mouth of the Gironde and that of the Seudre, the chain of dunes rises gradually in a northerly direction, and it is at the northern extremity that the highest hillock is found. This phenomenon is explained by the frequency and intensity of the south-west wind which blows in these parts; in virtue of "the parallelogram of forces," it carries the sand further and higher than the winds from the west and north-west can.

Every isolated dune assumes clearly-defined contours, resembling those of a crescent. It is easily understood why the hill must advance in such a manner as to project a curved point on each side of its principal mass. The grains of sand which the wind causes to ascend the height of the central part of the dune have to describe a longer path and to slide further down the counter-slope than the particles of the two lateral extremities. They proceed consequently with less speed; the ends, exceeding in rapidity the rest of the dune, bend forward, in the shape of advanced horns, and give the whole of the moving hill the aspect of a volcano whose crater has fallen in. That which contributes still more to cause these sandy hillocks to assume this semi-circular form is, that the prevailing wind does not always blow perpendicularly to the mass of the dune. Its direction is often oblique; now in one direction, and now in the other. It then makes the wings of the dune, the crest of which it strikes at right angles, advance more rapidly.

In the desert of Atacama, the Pampas of Tamarugal, in the plains of Texas, in the Sahara of Algiers, in the Nubian deserts, and in almost all the regions traversed by shifting sands, the crescent-shaped dunes present such a regularity of form that all travellers have been struck by it.[*] The Landes of Gascony also offer remarkable examples of this semicircular arrangement of the crest of the dunes. In the environs of Arcachon and La Teste all these hillocks have the appearance of fallen-in volcanoes, and are distinguished by the rich vegetation of broom and bushes which fill their craters or *crouhots*. In those parts of the coast of the Landes where the crater-shaped

[*] Poeppig, Meyen, Bollaert, Gillis, Laurent, Georges Pouchet.

form of the dunes is obliterated, it is evidently because two or more hillocks have been united, and, so to say, amalgamated by the im-

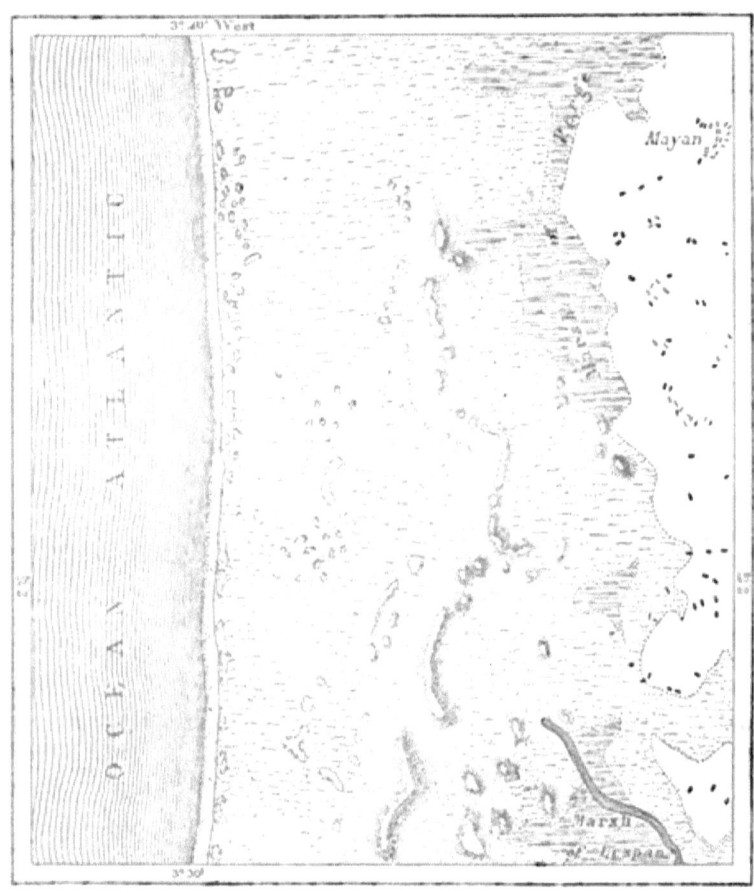

Fig. 92.—Crescent-shaped Dunes.

petuous wind which blows from the sea. However, we can account to ourselves for all these phenomena by studying the little swellings of sand, or miniature dunes, which are formed in thousands on the marine shores.

CHAPTER XXIV.

HEIGHT OF THE HILLOCKS.—ADVANCE OF THE DUNES.—DISPLACEMENT OF
"ETANGS."—DISAPPEARANCE OF VILLAGES.

In Europe the highest hillocks of sand are found on the coast-line of the Netherlands, on the Atlantic coasts of France, and in Scotland on the shores of the Firth of Tay. As to the dunes of the Mediterranean, they are generally lower than those on the coast of the ocean. The gulfs of the south of Europe having a hardly perceptible tide, their sandy shores are not incessantly displaced, like those on the strands of the ocean, and consequenctly they are less exposed to the winds which drive before them the finest particles. It is to the north of Africa, round the gulfs of the Syrtes, where the ebb and flow have the greatest development, and where sandy beaches occupy vast tracts, that the Mediterranean dunes attain the most considerable height. In France, those that are seen from Port Vendres to the mouths of the Rhone, hardly rise to more than 18 or 21 feet in height, because the banks on which these hillocks are formed have not a sufficient breadth, and above all, because the prevailing wind, the mistral, blows from the northwest, and carries the sand from the *étangs* into the Mediterranean.

On the coast of the Landes of Gascony, where the waves of the sea bring six millions of cubic yards of sand each year,[*] a great many dunes exceed the elevation of 225 feet. There is even one, that of Lascours, whose long ridge, parallel to the sea-shore, attains 261 feet in several places, and raises its culminating dome to a height of 291 feet. It is true that this height seems to mark in France the extreme limit of the ascent of the sand, for the ranges of dunes situated to the east of the dune of Lascours, are far less elevated. One would be tempted to admit that, after having arrived at this great height, the lower strata of wind from the west, compressed by the more elevated masses of air, have not the necessary power to cause the particles of sand to mount again, and are

[*] Laval, *Annales des Ponts-et-Chaussées*, 1842.

obliged to descend towards the plains of the interior, taking the crests from the hills previously formed. In Africa, on the low shores where the ocean bathes the great desert of Sahara, the enormous quantity of sandy materials that the eastern winds bring from the desert, and which the west wind drives back to the interior, permit, it is said, the dunes of Cape Bojador and Cape Verde to attain an elevation of from 390 to nearly 600 feet.*

The highest dune in the New World is perhaps that of Morro-Melancia, near Cape St. Roch, nearly 150 feet high; it rests on one side against a wooded hillock.

To the eyes of a traveller accustomed to the ascent of the Alps and the Pyrenees, these are very humble summits; yet these heights of sand assume the aspect of actual mountains, and their chains, arranged parallel to the shore, like ranges of enormous waves, seem to constitute an entire orographical system. Their bold taluses, their solid ridges, cut as with a chisel, the regular form of their tops, the general harmony of their contours, unceasingly varied at the will of the wind, give them an astonishing appearance of grandeur. The very even base-line which the sea-shore presents likewise aids to the illusion by contrast, and contributes to the grand aspect of these white hills. The old name, at once Celtic and Latin, of the dunes (dun), which was applied to mountains and steep hills, and which we still find in the names of several towns — Verdun, Loudun, Issoudun, Saverdun, proves that our ancestors had been singularly struck with the bold form of the sandy hillocks of the coast.

While gaining incessantly on the plains of the interior, the dune buries, without destroying, all solid objects, stones, rocks, trunks of trees, or human dwellings. Sometimes even it entirely covers pools of water, and causes them to disappear for some time under its sloping talus. When the sand brought by the wind falls regularly on a sheet of water, stagnant or covered with scum, it often forms a fine layer, completely veiling the water which bears it, from view. This bed can become solid enough to remain in equilibrium even when the level of the sea falls below it, and soon the particles of sand, dried by the solar rays, no longer betray the existence of the hidden pitfall. The herdsman or animals which set foot on the surface of the *blouse* are suddenly engulfed more or less deeply, and the waters of the pool rise around them. Most frequently they escape with the fright. Little by little the crumbling sand is heaped up; they allow the bottom to be consolidated, then

* Carl Ritter, *Afrika*.

quietly raising one leg, they wait till a sort of step is formed, and thus mount from stair to stair.

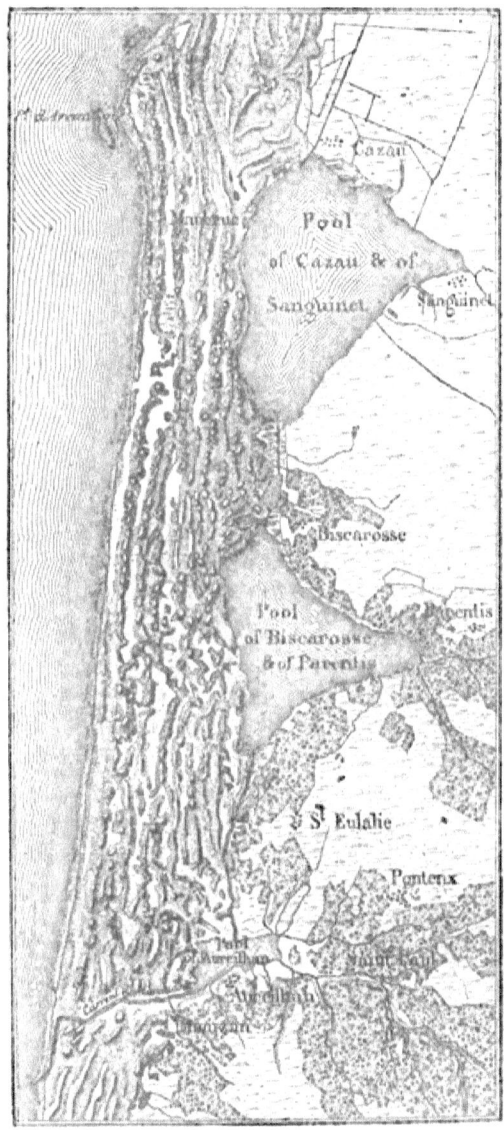

Fig. 93.—"Etangs" or littoral lakes of Cazau, Parentis, and Aureilhan.

If little pools are sometimes apparently swallowed, the more considerable masses of water, situated at the base of the dunes, are continually driven back into the interior. The rivers, arrested in their course and changed into marshes, are also forced to retreat, and mix their waters with those of the pools. This formation of lakes and marshes, parallel to that of the dunes, is one of the most remarkable features of the coast-line of the French Landes. A row of ponds, differing in form and size, but all situated at a nearly equal distance from the sea, is prolonged over a space of 125 miles. One large bay, the basin of Arcachon, has been able to maintain a wide communication with the ocean, owing perhaps to the river which it receives from the interior. But all the other sheets of water, to the north the étangs of Hourtin and Lacanau, and to the south those of Cazau, Parentis, Aureilhan, St Julien, Leon, and Soustons, only communicate with the sea by tortuous and rapid streams, and are now at a level considerably higher than that of the sea.

The étang of Cazau, the most elevated of all, and that which has been driven gradually inland by the strongest dunes, spreads its sheet of water at an altitude varying from 63 to 66 feet, according to the seasons. It has not less than 14,826 acres of mean superficies. The spectator who contemplates it from the top of a hillock would think he saw a vast marine bay, for a great part of the opposite shores escape the eye, and the isolated trees which mark afar off the distant bank, resemble a fleet of ships at anchor in a road; the white boulders of sand, of a triangular form, which are perceived at the foot of the green dunes, and which appear like so many sails of ships skimming along the coast, increase the illusion. Nevertheless, it is probable that the étang of Cazau was formerly a gulf of the ocean, for the bottom of this small inland sea is still found to be 36 feet below the marine level. The fishermen, who are the best authorities in such matters, uniformly attest, that in the lowest parts of the pond, the lead touches the sand at 15 fathoms. They also assert that it formerly communicated by deep trenches with the sea, and they even indicate the bay of Maubrucq as having been the ancient port, and trace the direction followed by the channel in the middle of the dunes. In the same way, the fishermen of the étang of Hourtin still show the site of the old port of Anchise.

It is easy to explain the gradual transformation of the ancient Gulf of Cazau, and other marine bays which indented the now uniform coast of the Landes. Separated from the ocean at first by a slender ridge of sand, as is often formed on low beaches, these bays which are changed into ponds, have been gradually driven inland by the parallel rows of dunes. Under the enormous pressure of the sand they have climbed, so to say, the slope of the continent. At the

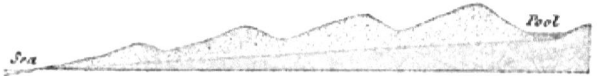

Fig. 91.—Formation of "Etangs."

same time the rains and rivulets, arrested in their course, have incessantly brought their contribution of fresh water to the new lakes, while the salt water retreated gradually by natural channels between the hillocks. Thus the grains of sand which the wind drives before it have sufficed, in the course of centuries, to change gulfs of salt water into ponds of fresh water, and carry them into the interior of the continent, to a height considerably above the Atlantic.

The same phenomena occur also in the sandy islands which are found in the middle of the sea. The greater part of these islands have a perfectly regular form, due at the same time to the currents which bathe them, and to the winds which form the dunes. In

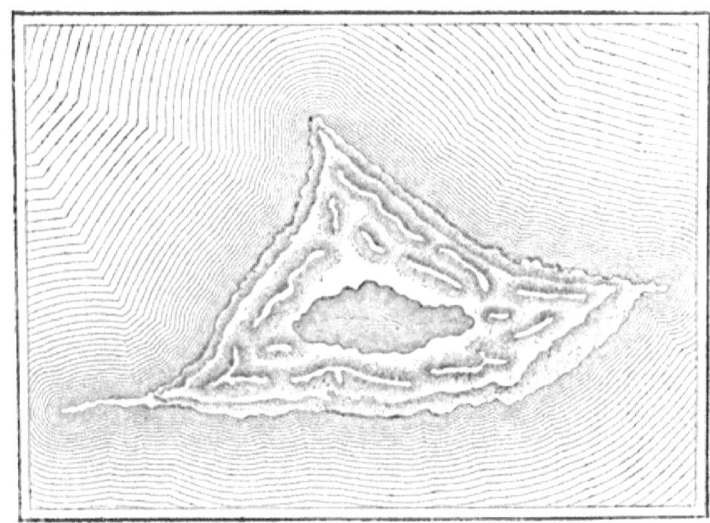

Fig. 95.—Isle Thelenji in the Caspian Sea.

the centre of the triangular or crescent-shaped space which they surround with their moving hillocks, they enclose one or several ponds, which formerly were a part of the sea, and which have transformed themselves by degrees into pools of saltish and finally fresh water. In Sable Island, situated not far from the mouth of the St Law-

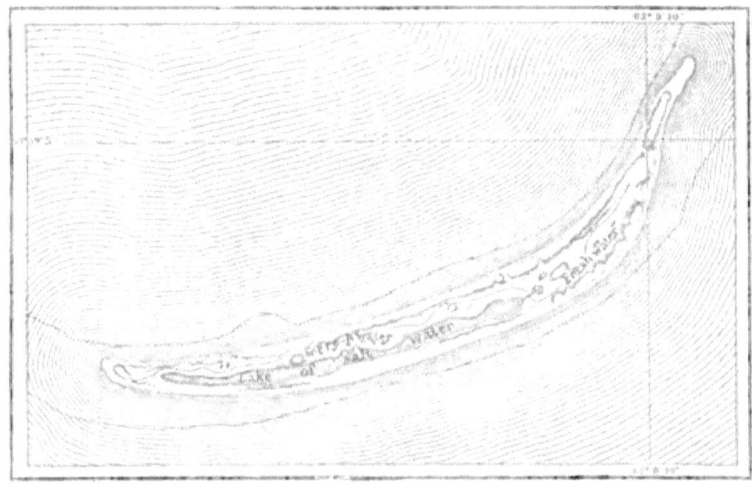

Fig. 93.—Sable Island.

rence, we can observe this phenomenon of transition actually in progress. While the large lagune of the interior, too extensive to be rapidly purified, is still filled with salt water, the small ponds lying between the dunes are already fresh water.

Numerous have been the disasters occasioned by the invasion of dunes or ponds during the historic era. The villages situated at the eastern base of the dunes of Gascony, on the shores of the ponds, must be moved from time to time towards the east, for fear that they should be swallowed by the sands or the waters. At the approach of the danger the threatened inhabitants sometimes attempted a vain resistance. As soon as an east wind succeeded to the regular winds from the west, herdsmen and labourers, armed with spades and pickaxes, repaired in all haste to the top of the dunes, and, filled with a purposeless ardour, they destroyed the crest of sand, and delivered it to the power of the wind. But the regular winds soon brought back the sand again, the dunes recommenced their advance, and routed the army of peasants. For fear of being buried they were obliged to destroy their huts, in order to carry away the materials and build new dwellings at a certain distance inland. Years and centuries passed; but the dunes and the ponds constantly advanced, and the inhabitants were again condemned to transport their villages into the midst of the heaths. These were foreseen misfortunes, and the chronicle preserves silence as to the successive emigrations; it confines itself to mentioning the names of some churches which were obliged to be abandoned to the sands and reconstructed far on the plateau of the Landes. Thus we know that the church of Lége was re-built in 1480 and in 1650, the first time at $2\frac{1}{2}$ miles, the second at nearly 2 miles further inland; but the halting-places of other local monuments of the same district are not known in an exact manner. As to the now vanished towns of Lislan, Lélos, and many others, their ancient situation is unknown. After having lost its port and its hamlets, the township of Mimizan, formerly very important, was about to be entirely buried when, at the last moment, they succeeded in fixing the dunes by palisades and plantations. The semicircle of invading hills, like the serrated mouth of a crater, still seems to be on the point of devouring the houses.

Dunes have often been compared to gigantic sand-glasses measuring time by the progressive march of their sandy talus. The comparison is just, for the western winds, which effect all the changes on the coast-line of the Landes, obey at present the same laws as they did thousands of years ago, and very probably their force has not changed during that interval of time. The dunes, the ponds, and even the villages on the shores, may thus be considered as real geological chronometers; but, unfortunately, the indications that they furnish have not yet been deciphered with any certainty, and now that the dunes are fixed, it is too late to undertake this study. The illustrious Brémontier (whose book, printed in the year 1797,[*] is still an authority on the question of moving sands) collected during eight years a series of observations which have given an average of from 22 to 27 yards for the annual progress of the dunes of La Teste. This result agrees in a remarkable manner with the indications furnished by the encroachments of the dunes of Lége during the last four hundred years. In admitting as normal the average calculated by Brémontier, one would arrive at this conclusion, that in the lapse of twenty centuries the dune would be able to invade the entire district of the Landes and cover the town of Bordeaux. A thousand years would even have sufficed to transform the fair plains of Bordelais into marshes, for the *étangs* constantly driven back by the invading dunes, would have spread on the eastern side after having

[*] *Mémoire sur les Dunes.*

passed the culminating line of the plateau of the Landes. Researches undertaken in other places would have doubtless confirmed the observations made by Brémontier; but in the absence of these, we cannot accept measures taken at the foot of a group of isolated dunes as applying to the host of sands from Bayonne to the Point de Grave. In order to pronounce definitely, we must wait for observations which will not fail to be made one day, on the advance of the dunes in all those parts of the globe where these hillocks have not yet been arrested.

CHAPTER XXV.

OBSTACLES OPPOSED BY NATURE TO THE PROGRESS OF DUNES.—FIXATION OF
THE SANDS BY SEEDS.

THE work of nature is, however, double; and if on one side she
hastens the advance of the sands, on the other she attempts to arrest
them. She herself points out the means of prevention, or else prevents spontaneously the disasters of which she is the cause. In certain
places, and especially on a part of the coasts of the Landes, she
exercises a physical and chemical action by employing the oxide of
iron which the water contains to consolidate the sands and transform
them gradually into actual rocks. Elsewhere organic cements, composed of broken shells and remains of silicious and calcareous infusoria,
agglutinate the arenaceous particles, and give them the necessary stability to resist the winds. But these means of consolidating the
sands are exceptional. It is principally vegetation which fixes the
moving hills on the sea-shore. On almost all coasts the sandy and
calcareous *débris* of the soil contain enough fertilizing principles to
nourish a certain number of hardy plants, which do not fear the salt
air of the sea, and which send down their roots to a great depth, so
as to absorb the necessary moisture. Among these hardy vegetables the commonest, and most useful at the same time, is Marram-
grass (*Arundo arenaria*), whose slender and flexible stems can hardly
arrest the wind, but whose strong roots, sometimes 12 or 15 yards
long, develop all the better the less consistence the sand has.
Various species of convolvuli creep over the ground, and fixing
their vigorous cordage from place to place, sometimes envelop an
entire dune in their network of leaves and flowers. Other plants
rise more proudly, but if their stem is buried in the sands they transform it into a root, and give birth to a new shoot, which may be interred in its turn, without the plant being in danger of perishing.
Thus such a seed germinating at the base of the dune often produces a plant which ends by spreading to the summit of the mountain, and fastens by a cable of roots the arenaceous strata which the

creepers of the convolvulus fix on the surface. A number of plants, whose frail stems are half buried in the sand, are, perhaps, contemporary with the dune itself;[*] perhaps even they existed before mankind had a history.

In this strife between the force of the winds and the power of vegetation, the definite issue depends at the same time on the climatological conditions, the nature of the soil, the form of the shore, and various other circumstances, among which we must rank, in the first place, the havoc caused by men and animals. In South America, on the shores of those tropical countries where the development of plants is favoured, according to the seasons, by an extreme heat and by torrents of rain, and where the sands contain a considerable proportion of animal and vegetable remains, most of the dunes are already fixed at a few yards from the sea by mimosas, cactuses, and thorny trees. However, on the eastern shores of all the rivers of equatorial Brazil, which discharge themselves near the mouths of the Amazons, we see, even somewhat far from the sea, ranges of dunes from 25 to 50 feet in height, which move incessantly, driven by the breezes of the trade winds.[†] This mobility of the sands is undoubtedly connected with this fact, established beyond question by Coutinho and Agassiz, that the shores are depressed in that part of Brazil, and, consequently, they incessantly change their form, so that the dunes have not yet had time to be fixed.

In Europe the flora of the sands is less rich than in equatorial countries. On the coasts of Jutland it is composed of only 234 species of plants, very insignificant for the most part,[‡] and the "white" dunes of the Danish peninsula, as well as those of Gascony and Holland, have also not enough cohesion to resist the furious western winds which assail them. It is probable, nevertheless, that even in the countries of the temperate zone the modest herbaceous vegetation of the sands of the coast could, after a certain lapse of centuries, acquire the strength necessary to fix the dunes, and prepare, by the slow accumulation of its remains, a vegetable bed, where large trees would grow spontaneously.

If it were not so, it would be difficult to understand how all the dunes of Europe were originally covered with forests. According to the unanimous testimony of the ancient geographers, the woods extended to the sea-shore in those plains which are now the Nether-

---

[*] Aug. Pyr. de Candolle; Elie de Beaumont.
[†] *Revue Maritime et Coloniale*, 1866.
[‡] Andresen, *Om Klitformationen*; Marsh.

lands, and the Batavians, the Angles, and the Frisons had no special word in their idioms which designated a hillock of moving sand.* Neither the great geographer Strabo, nor Pliny, the encyclopedist, nor any other writer of antiquity, mentions the existence of hills driven by the wind, though this phenomenon was certainly of a nature to strike them. Under a great many of the dunes of Gascony trunks of oak and pine trees, with other substances, are discovered buried in the sand, above the ancient level of the Landes. Moreover, some dunes still bear magnificent woods, which can count at least several centuries of existence, and which probably were not planted by man. Not far from Arcachon one may wander in a forest where gigantic pines rise, unrivalled in France, and oaks 46 feet in circumference. Title-deeds of 1332 speak also of forests which covered the dunes of Medoc, and where the seigneurs of Lesparre went in merry company to chase the stag, the boar, or the roebuck. Montaigne,† too, writing in the middle of the sixteenth century, says that invasions of the sand had taken place "for some time." Besides, why should the Landese, like the Spaniards, give the name of *monts* or *montagnes* to their forests, even those of the plains, if not because their hills of sand were, in former times, uniformly covered with trees?

Unhappily all those fine forests which once protected the low lands of the sea-coast against the invasion of the sands, were successively destroyed during the evil days of the Middle Ages, either by barbarian invaders, or by improvident lords, or by the peasants themselves. Even in the last century the King of Prussia, Frederick William I., being in great want of money, caused the forest of pines to be cut down, which extended without interruption over the dunes of the Frische Nehrung from Dantzig to Pillau. The operation brought him the sum of 200,000 crowns, but the moving sands invaded the great inland bay, destroyed the fisheries, obstructed the navigable channel, buried the defending fortresses, and modified in the most vexatious manner the hydrographic economy of all those parts.‡ In Holland and in Brittany this dismantling of the coast has produced still more fatal results. On the borders of Lake Michigan, and at Cape Cod (Massachusetts) the clearing of the shore has also produced the formation of moving hills.§ But the inhabitants have only themselves to complain of; the dunes are their work.

---

* Staring, *Voormals en Thans*; Marsh.   † *Essais*, livre iv.
‡ Foss, *Zeitschrift für die Erdkunde*, 1861.
§ Marsh, *Man and Nature*.

A single imprudence may cause great misfortunes; and thus, according to Staring, one of the highest dunes of Friesland owes its origin to the destruction of a single oak.*

It is for man now to arrest by his labour those hillocks of sand which he has, so to say, created by his imprudence. Happily this is not an impossible task. The shepherd of the French Landes, when he wishes to protect his cabin, erected in the depth of some ravine of the dunes, takes care to cut, in the lédes and surrounding marshes, grass or rushes, which he spreads over the soil in such a manner as to cover it completely, and to leave none exposed to the sea-breezes. This is sufficient; the sand remains immovable, and the dune is fixed for the future; so long at least as no horse's foot, or the teeth of a sheep or wild animal, a shower of rain, or any other cause, have penetrated the protecting layer and restored their mobility to the sands. It is then necessary to carpet the ground with a new litter of plants.

This means of protection, which is moreover only practicable over small extents, is evidently quite provisional; to obtain a definite result we must have recourse to the direct fixation of the dunes by the seeds of trees or other plants, so as to present an insurmountable barrier to the winds. In modern times the Dutch, those great masters for all works concerning the sea and the coasts, have been the first to recognize the absolute necessity of arresting the dunes. Defended and menaced at the same time by those masses of moving sand which never cease to encroach on their territory, even while protecting it against the assaults of the sea, they have understood that the very safety of their country may depend on this rampart of hills, and for a century they have effectually consolidated it by planting reeds, maples, and firs.

The first attempts at the fixation of the dunes made in Gascony date from the beginning of the eighteenth century. M. de Ruhat, who had acquired the ancient Captalate de Buch, sowed some of the hills of La Teste with pine trees; but though this plantation succeeded perfectly, the work was not continued, and everywhere else the indolent Landese allowed the dunes to advance to the assault of their villages. Later, the brothers Desbiey and the engineer Villers proposed repeatedly, at various times, the fixation of the entire district of sands. Their voices were not heard. It is to the celebrated Brémontier that the honour is due of first causing to be adopted and put in practice a complete plan for the culture of all the dunes. Inspired with the writ-

* *De Bodem van Nederland*, tome i. p. 425.

ings and example of his predecessors, and not disdaining to interrogate the herdsmen, who knew by tradition the means of arresting the sands, Brémontier first applied himself to the task in 1787. The works were interrupted in 1789, resumed in 1791, and completely abandoned again in 1793, in consequence of the opposition given by several of the inhabitants of La Teste. But important results had been already obtained. More than 620 acres of moving sands had been fixed in the environs of Arcachon; pines, oaks, and vines were in perfect growth, and the sowing of every two acres had not cost more than 200 francs. The possibility of arresting the advance of the dunes at little cost was perfectly demonstrated.

At the commencement of this century the interrupted work was resumed, and it was completed some years ago. The dunes of Gascony, fixed for the future, enrich the countries which they formerly threatened to bury, and in consequence of the increasing value of the pines and their productions, we must reckon the annual increase of public wealth on the coast at hundreds of thousands of francs. The estimated present value of the forests of the Landese dunes is 25 millions; that is to say, 600 francs the acre. Thus, the means of safety applied by Brémontier has become a cause of prosperity to the inhabitants. At the same time, several happy results, which could not be looked for at first, have been obtained. The sand, protected from the rays of the sun by the shade of the pines, produces herbs which are utilized as straw or food for cattle. The marshes, which during six months of the year were transformed by rainwater into impenetrable morasses, have been drained without the intervention of man, owing to the thousands of roots constantly pumping up the moisture from the sands. The surface of the vast ponds, situated at the eastern foot of the dunes, is lowered likewise to furnish the forest trees with the water necessary for their growth. Besides this, the fixation of the dunes has caused the "blouses" to disappear, in which men and animals were engulfed; the sands do not advance any further, and the pools have ceased to exist. Science has repaired the disorders formerly caused by man's imprudence.

PART II.

THE ATMOSPHERE AND METEOROLOGY.

BOOK I.—THE AIR AND WINDS.

CHAPTER I.

AIR THE AGENT OF THE VITAL CIRCULATION OF THE PLANET.—PHENOMENA OF REFLECTION AND REFRACTION.—MIRAGE.

DEATH and eternal silence would reign over all the earth if it were deprived of the atmosphere that envelops of our planet. This gaseous, transparent, and invisible mass, which scarcely seems to form a part of the earth, is nevertheless its principal element. For it is the most mobile, and it is by its agency that life is sustained. The earth supports us, but whether men, animals, or plants, we alike require the air for our existence. Although not flying in it like birds, all living beings, whether they walk, climb, or fix their roots in the soil, are not the less children of the atmosphere.

Considered as one of the heavenly bodies, our planet is composed of a solid kernel surrounded by two fluid strata. The kernel is that which bears more especially the name of the earth, it is the rocky beds containing lava, molten metals, and the entire mass of unknown substances which occupies the centre of the globe. The sheet of water forming the seas and the network of rivers covers this solid skeleton, and above this watery envelope is stretched a second spherical layer still more fluid, and whose currents and counter-currents incessantly circulate from the pole to the equator, and from the equator to the pole, with the regularity of the lungs of man, by turns filled and exhausted. The atmosphere is truly the breath of the planet; like its satellite, which most astronomers tell us is destitute of a gaseous envelope, the earth would be only a dead star rolling in space if it suddenly lost the stratum of air that surrounds it and ceased to respire the regular breath of the winds.

The subtle and transparent air is composed of the same gases which are found in greater abundance in the opaque and solid crust of our globe. The four principal elements of all vegetable or animal organism, oxygen, nitrogen, hydrogen, and carbon, are found likewise in the atmosphere. The two first as constituent elements of the air, the third united with oxygen under the form of watery vapour, and the fourth mixed with the breath expired by animals, and with many other gases resulting from the decomposition of organic matter. Between the action of nature and the eternal movement of the atmosphere an exchange is constantly being effected, by which the gases, one instant in the animal, plant, or rock, fixed in an organism or in the terrestrial strata, are disengaged and recompose the atmosphere.

Animals and plants would soon be all destroyed, for want of necessary aliment, if the mixture of vapours and gas were not effected by the incessant movement of the aerial masses. Men and animals would gradually kill themselves by absorbing again the carbonic acid already expelled from their lungs; and plants plunged in an atmosphere too full of oxygen emanating from their leaves would end too by dying. Happily, the currents of air, which pass in immense spirals over the surface of the earth, uniformly mix all the gases they carry away with them, and thus favour life over their whole course. To the temperate regions, which are principally the domain of man, they bring the oxygen which the immense forests of the tropical zone have exhaled; to these same forests they impart the carbon which is life to trees, and would be the death of man. Still more, they animate the globe itself, by carrying immense quantities of vapour to the mountains where the net-work of springs is elaborated, and in causing to circulate above the sea a dry air eager to absorb the water which evaporates from its surface. Like the heart in living organisms, the productive zone of the atmospheric currents occupies the central region in the ocean of air, and moves alternately to the north and south. It is thus that a movement of systole and diastole is produced in all the aerial mass, impressing the initiatory speed to the arterial currents which carry fertility to all points of the planet.

Every particle of gas passes thus continually from life to life, and escapes from death to death; by turns, wind, wave, earth, animal, or flower, despite its smallness, is the symbol of infinite motion. The air is an inexhaustible source whence all that lives, draws its existence, an immense reservoir into which all that dies pours its last breath. Under the action of the atmosphere all the scattered organisms are born and perish. Life and death are equally in the air which we breathe, and perpetually succeed one another by the exchange of gaseous particles. The same elements which are exhaled from the leaves of the tree are carried by the wind to the infant that is just born ; the last breath of a dying man goes to form the brilliant corolla of the flower, and compose its penetrating perfume. The breeze which gently caresses the stems of the plants is further on transformed into a tempest, uproots large trees, and destroys ships, with all their crews. It is thus, by an infinite series of minor catastrophes, that the atmosphere sustains the universal life of the globe.

Comparable to the ocean, as to the incessant circuit of its waves, the great atmospheric sea is not however enclosed in a basin bounded on all sides. The atmosphere travels without cessation, bearing away on its wings all the light objects which are exposed to its currents. It takes up the ashes from a crater in eruption, and lets them fall in places often hundreds of miles distant ; it raises in its eddies myriads of animalcula or clouds of pollen, which are wafted by it across seas to fall again in impalpable dust. It carries the sea itself in the form of clouds, and distributes it as rain and dews over the continents ; it becomes highly charged with electricity, and discharges itself by the rays of the Aurora Borealis, or by vivid lightnings. It is the great vehicle by means of which the universal interchange of the elements which compose the solid crust, the mass of waters, and of organic beings, is accomplished.

"The world is small!" said Columbus ; but it is principally owing to the air which disregards distances that the earth is diminished. Whatever be the number of yards or miles traversed by a seed, the point where it falls is not distant from the mother-plant. The northern coasts of the Mediterranean are brought nearer the great deserts of Africa, dust of which is brought by the *sirocco ;* and in the same way we may say that the shores of Brazil, towards which the trade-winds blow, are contiguous to the distant archipelagos of the Azores and the Canaries. All those parts of the world united by atmospheric currents become thereby neighbours to each other, if not for the creatures who walk on the ground, at least for those which are carried by the movements of the air. By the incessant mixture of the aerial masses all the regions of the solid kernel of the earth are brought nearer, contrasts are blended, and harmony is established between the productions and the climates, no less than in the general aspect of nature.

The winds are also powerful geological agents. Thus the aerial currents of certain latitudes transport clouds of dust, which may at length render vast countries sterile or fertile, either by covering the natural soil with an unfruitful layer, or by effecting a happy mixture with it. On the banks of the Nile, the sand of the desert which the wind mingles with the thick mud of the river, contributes to develop the marvellous productive force of the land, while in the neighbouring plains, which are destitute of moisture, it buries the plants and renders the soil wholly unfit for vegetation. Elsewhere, and principally on the low coasts of the sea, the wind drives hills of sand across the plains, barring the outlets of the streams, and gradually driving the water up the slope of the continent.*

In certain places the aerial current even goes so far as to temporarily change the level of the sea; it sometimes arrests the waves, or hurls them against the shores, and alternately dries up the bed, and causes disastrous inundations. Sometimes the wind, which descends with violence from the polar regions of North America to the Gulf of Mexico, keeps back three or even four successive tides. Then these, returning altogether in one foaming mass, sweep over whole islands off the low coasts of Louisiana and Texas. In the same way when the *pampero* or south-west wind blows over the great estuary of La Plata, its waters are sometimes lowered by 12 or even 18 feet in less than half a day, and the vessels that were floating in the road remain stranded in the mud.†

This is not all. The wind can also modify the configuration of the shores, since the waves of the sea, which contribute so largely towards the sculpturing of them, receive from it their impulsive force. Thus the large arm of the Rhône perhaps owes its south-easterly direction to the *mistral* which descends from the Cevennes.‡

As to the delta of the Mississippi, its exterior contours are probably modelled by the south-east monsoon which prevails in that country; the southern passage, which opens exactly in the direction of the prevailing wind, is almost entirely obstructed by the dike of mud that the surf has raised across its current. The two arms of the Mississippi which carry the greatest quantity of water are directed,

* See above, p. 211.
† Fitzroy, *Adventure and Beagle*, vol. ii. Appendix, p. 89.
‡ See *The Earth*, the section entitled, *Rivers*.

the one to the south-west, the other to the north-east; that is to say, each of them forms a right angle with the monsoon from the south-east. It is the aerial current which has forced the long peninsulas of the Mississippi to spread thus over the waters like the branches of a great fallen tree.*

The geological labours of the winds are, however, accomplished, for the most part, in an indirect manner, either by the evaporation of the moisture of the continents, or by causing considerable downfalls of rain. During the course of ages the contours of the land and sea have not ceased to change, and, in consequence of these gradual modifications, the winds themselves have been subjected to analogous variations. Some are saturated with the vapour of water, and the clouds that they carry are deposited in rivers and lakes in the midst of land. Other atmospheric currents have lost their moisture in great part, and then in passing over inland seas they have absorbed them; pumped them, so to speak, leaving behind them smiling plains transformed into deserts. Without any doubt it is the winds which have now dried the lands of Cape Natal and Transvaal; it is they that have been the great agents in the work of drying up central Asia, they have drunk the vast extent of water that formerly stretched from the Euxine to the Caspian Sea, and from the Lake of Aral to the Gulf of Obi, and left steppes of salt in the place of this ancient Mediterranean.†

It is by means of the atmosphere, too, that the exchange of particles between the earth and bodies wandering in space is accomplished. When an aërolite, shot like an enormous bullet through space, meets the exterior strata of gas that surround the earth, it is instantly set on fire, and bursts either entirely or on the surface; and hurling with violence some fragments to the ground, it leaves behind it a long train of luminous matter resembling a fiery track. Owing to the resistance opposed by the atmosphere to the passage of the strange star, the globe is every year enriched in this manner with material brought from the sky. The strata of air, moreover, are the vehicle of all sounds; they also convey the vibrations of light and heat. Deprived of this envelope the globe would immediately be wrapt in complete darkness. But if the atmosphere allows the rays of luminous heat, emitted by the sun, to pass, it intercepts in return a great part of the dark rays, which escape from the earth into space. It is thus that

---

* Humphreys and Abbot, *Report on the Mississippi River*, p. 450.
† Maury, *Physical Geography of the Sea.*

the globe has been able to preserve its normal temperature, and has become the theatre of life.*

The atmosphere, which as the common vehicle of exchange is ever in motion, is also the great agent by which nature receives the wonderful colours that beautify her. It is owing to the reflection of the blue rays that the sky and the distant heights of the horizon assume that beautiful azure hue, which varies with the altitude of each region, the abundance of watery vapour, and the contrast of the clouds. It is owing to the refraction undergone by the luminous rays in passing obliquely through the aerial strata, that the sun is announced every morning by the vague glimmers of twilight, then by the splendours of dawn, and thus shows himself before the astronomical hour of his rising. It is also due to an analogous phenomenon that in the evening he seems to slacken his descent below the horizon, and even after he has disappeared colours the west for a long time with the purple of sunset. Without the gaseous envelope of the earth we should never see those varied plays of light, those changing harmonies of colour, those gradual transformations of delicate shades, which form the marvellous beauty of our mornings and evenings. The special works on meteorology describe at length all these brilliant phenomena of the air, the rainbows, halos, parhelions, and that splendid spectacle of the "after-glow" which colours the snows and ice of the Alps with a rosy tint more than twenty minutes after the sun has set. Nothing is so beautiful as this phenomenon due to the contrast of the lower slopes which are already in the shade, and the high peaks which the solar rays reflected above the horizon still illuminate. When the Aiguille-Verte is already veiled in shadow, as well as the neighbouring summits of Mont Blanc, the latter is truly transfigured by the light glittering on its snows. "We might think we then saw a form foreign to the earth;" then all at once the flame is extinguished, the colours so brilliant vanish, "to give place to an aspect that we may truly call cadaverous, for nothing approaches more nearly to the contrast between life and death on the human face than this passing from the light of day to the shadow of night on the high mountains." †

The *mirage* is another singular optical effect, due to the deviation of the rays of light which traverse the atmosphere. When the surface of the earth is much heated by the sun the lower strata of air expand and often become lighter than the strata situated above. If

* Tyndall, *Heat*.
† Naker de Saussure, *Annales de Chimie et de Physique*, 1839.

the air is agitated by the wind it then rises oscillating like the smoke that rises from a high furnace, and the outlines of all objects seen through this vapour seem to tremble; if a calm reigns in the atmosphere all the objects bathed by the denser strata are reflected, as in a sheet of water, in the more expanded air, and

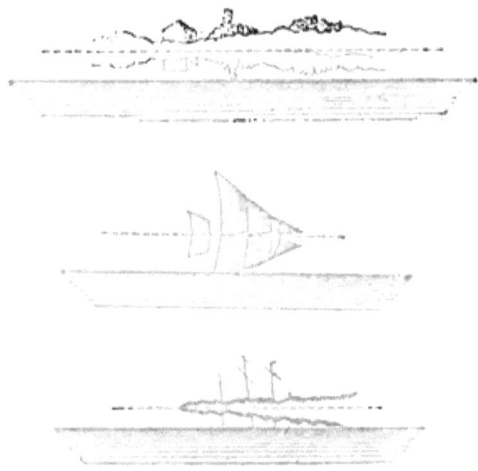

Fig. 97.—Mirages at Verdon, at the mouth of the Gironde.

all their images appear double; hence the name of *espeio* (mirror) which the inhabitants of South America give to the mirage. In the midst of the arid desert, at hundreds of miles from any stream, bushes and rocks are reflected in the air as in the basin of a fountain; on the sea the ships, the shores, and signals are reproduced as on a second ocean, even in the large squares of our cities which a burning sun strikes, the statues sometimes seem to bathe their feet in a crystalline water reflecting their graceful forms. This optical illusion which thus paints imaginary objects even in our cities, is the "Fata Morgana" of Italy, the deceptive "Delibab" of the Magyar puszta, and the "Thirst of the gazelle" on the plains of Hindostan. It shows from afar fresh oases and rippling waters to the fatigued travellers, who, where the deceitful picture glitters, only find aridity, thirst, and perhaps death. In the deserts of Arabia the plain seems every day transformed into an immense lake. In proportion as the sun sinks the magic sheet retires, then it fades completely away, to reappear the next day an hour or two before noon.*

* Palgrave, *A One Year's Travel in Central Arabia.*

The phenomenon of reflection is almost always accompanied by lateral movements which apparently alter the position of objects, in the same way as plates of glass of unequal thickness do, we then see large masses of different forms detaching themselves to the right and left of the distant objects, and floating fantastically in the air. These phenomena of mirage are most curious in the polar seas already strewn with blocks and icebergs of every variety of contour. The surface of the ocean bristles with points, needles, crests, and overhanging cornices, which separate, rejoin each other, and then vanish, to reappear again. Nowhere do we see more astonishing phantasmagoria. As to the prodigious scenes that the mirage is said to present to the eyes of the traveller, by showing him forests of palm trees, temples with

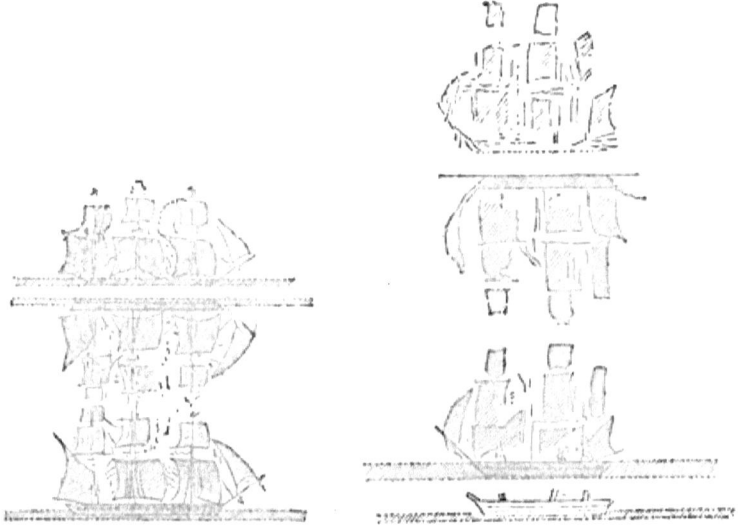

Fig. 93.—Mirages of the "Vincennes" and of the "Peacock," after Wilkes.

colonnades, caravans, armies on the march, and people gathered for fêtes, they are probably in great part produced by fever under the ardent sun, for in this fiery atmosphere which floats above the whitened plains and reflects the splendour and the heat, the head is burning, the imagination excited, and the eye sees no longer anything but the forms of fancy.

CHAPTER II.

WEIGHT OF THE AIR.—HEIGHT OF THE UPPER STRATA.—BAROMETRIC MEASURES.

The weight of the aerial particles, which makes itself felt in so terrible a manner in hurricanes, is relatively very small, since a litre of air (nine-tenths of a quart) taken at the surface of the ground, and at the temperature of zero, weighs 770 times less than a litre of water. But the atmospheric mass surrounding the globe is such that if it were to be entirely agglomerated in a single ball, it would weigh as much as a sphere of copper nearly 63 miles in diameter; that is, the twelve hundred thousandth part of the mass of the earth.* The pressure exercised by the atmosphere on a man of middle size is not less than 14 or 15 tons; it is true, however, that this pressure making itself felt at the same time in all directions on our frame, is by that very fact neutralized. We know that a column of air on any point whatever of the earth is equivalent on an average to that of a column of water of 32 feet, or to 30 inches of mercury; it is the knowledge of this fact that has enabled us to construct the barometer.

Still if we know the weight of the atmosphere, we cannot yet say in a positive manner to what distance it rises in space. If the higher aerial strata had the same density as those on the surface of the sea, the total thickness of the air would not exceed 5 miles, and consequently, the highest mountains of the earth, the Gaourisankar, the Kinchinjunga, the Dapsang, and many others, would raise their peaks into empty space above the atmosphere. But it is not so; above the lower strata, compressed by the weight of all the superincumbent aerial mass, the particles separate in proportion as the pressure diminishes, the air becomes rarer and rarer in the heights of space, and ends by being completely lost like the thin fluid which composes the tail of comets. According to the calculations of Laplace, it is at more than 26,000 miles above the surface of the earth that, in consequence of the increase of centrifugal force, and the diminution

* Sir John Herschel, *Meteorology*, p. 16.

of the weight, the aerial particles which may still be in space must forcibly escape from the terrestrial orbit. Perhaps it is, in fact, in these elevated regions at the very limits of the spheres of attraction of the heavenly bodies that the exchange of their gaseous particles takes place. However that may be, it is at a height very inconsiderable in comparison with the extreme limit indicated by Laplace, that the atmosphere ceases to be respirable by man. At the summit of Etna, that is to say, at an elevation of two miles, we have nearly a third of the aerial mass under our feet. At $3\frac{1}{2}$ miles, a height above which a great many mountains raise their peaks, the column of air which rests on the ground has already lost one half of its weight; consequently, all the gaseous mass which extends far into the sky to immeasurable distances, is simply equal to the aerial strata compressed into this lower region.

More than two hundred years ago, Perier, following the indications of his brother-in-law Pascal, established by the first direct experiment the diminution of the weight of the air in a vertical direction; he ascended Puy-de-Dôme, with the barometer in his hand, and during the ascent the column of mercury which measured the atmospheric pressure never ceased to sink gradually in the tube, and thus the means of measuring the height of mountains above the level of the sea, by simply reading the barometrical indications, was discovered. Since this epoch science has made great progress, the precise law of the decrease of the weight of the air and all other elastic gases has been brought to light by Mariotte, and innumerable travellers have been able, with the aid of the barometer, to indicate approximately the altitude of the salient points in the various countries that they have traversed. Nevertheless, one can never be sure that the barometer has furnished perfectly exact measures of height. In each barometic reading we must take into account the temperature, the quantity of watery vapour contained in the atmosphere, the agitation of the winds, in a word, all those physical conditions of the air, whose weight we are about to measure, and each of these secondary observations makes a greater or less correction necessary in the final result. The direct results obtained by trigonometry are at present the only ones that give in an exact manner the height of the surface.

To ascertain the altitude of summits another means is also employed, which, in consequence of the defectiveness of the instruments, generally gives results still less exact than those of the barometer. This means consists in measuring the heat of boiling water. In fact, the boiling point, or the temperature at which the tension of the vapour of water exactly balances the atmospheric pressure, must necessarily sink in proportion as the pressure diminishes. It has been calculated that the average fall of boiling point is 18° Fahrenheit for every 1062 feet of vertical height. But experiments may give for the heights of mountains differences of many hundred feet. Thus, Tyndall found in August, 1859, that the temperature of boiling water on the summit of Mont Blanc was 84·97°, while in the preceding year he had observed a slightly lower boiling point on Mont Rosa, though this latter peak is 558 feet lower than the giant of the Alps.

To what height is the air dense enough for a man to be able to find the oxygen necessary for his lungs, and to live there for a few seconds at least? The climbers of mountains have never reached this extreme limit, because of the fatigues of the ascent, which add to their difficulty of finding a sufficient quantity of air. Thus the highest peaks of the Himalayas and the Andes have remained to this day untrodden by human foot.* At the summit of Ibi-Gamin, the highest point yet scaled in an ascent, Robert Schlagintweit found himself at an elevation of $4\frac{1}{4}$ miles. The barometer was only 13·3 inches, so that the traveller had beneath him nearly three-fifths of the mass of air.

Nevertheless, thanks to the balloon, aeronauts have been able to ascend to heights, which even the condor does not reach, and from whence the highest mountains would appear as if they rose from the depth of an abyss. In 1804, Gay Lussac ascended to $4\frac{1}{2}$ miles; in 1851, Barral and Bixio ascended a little higher; in 1858, Rush and Green rose to 5 miles. But these are all altitudes inferior to the highest summits of the continents. Finally, on September 5, 1862, Glaisher and Coxwell undertook an aeronautic expedition, in which they resolved to ascend as long as they could preserve the sense of their own existence. The air becoming too rare for their lungs, hardly allowed them to pant, they had palpitations of the heart, singing in the ears, the blood swelled the arteries of their temples, their fingers froze and refused to move; but their will sustained them, they threw more sand from the car, and thus gave themselves a new impetus into the atmosphere. Glaisher fainted away, but his companion did nothing to arrest the ascent; his eyes fixed on the instruments, he noted with a glance the gradual sinking of the column of mercury in the barometer and thermometer, as if he were in the observatory at Kew. Gradually taken possession of by torpor, the aeronaut lost the use of his hands; but he still held

* See *The Earth*, the section entitled, *Mountains*.

the cord of the valve between his teeth, and when he felt that but one single second separated him and his friend from death, then he let the gas escape, and the balloon was arrested and descended gradually towards the plains situated at $6\frac{1}{2}$, or perhaps at $6\frac{3}{4}$, miles below, for the column of the barometer was only at 6·5 inches. What noble courage on the part of these men risking death with such simplicity of soul, and for the sole advantage of studying the temperature of an atmosphere where neither man nor bird can live! Certainly, it would greatly lower the force of soul and philosophical calm to compare it to the brutal courage of the soldier rushing into the thickest of the furious mêlée, intoxicated with powder, din, and blood.

At the height to which Glaisher and Coxwell rose, they had nearly four-fifths of the weight of the atmospheric strata beneath them; the remaining fifth, where the air is too rarefied for the lungs of man, rises dilated more and more to unknown heights. We can, however, ascertain the presence of the aerial fluid much above the space to which man has been able to ascend. In truth, the refraction of the solar rays in the dawn and twilight has permitted us long ago to calculate that the appreciable part of the atmosphere rises to, at least, 45 miles, and owing to the perfection of optical instruments, the visible limits of this ocean of air, which bathes our globe, have gradually retreated. In supporting the observations made in tropical regions on the phenomena of the twilight, M. Emmanuel Liais believed he could affirm that the height is in reality 200 and even 210 miles.*

By this the real diameter of the earth would be increased by about a tenth. Though this atmospheric stratum is usually left out of the calculations of astronomers on the dimensions of the planet, it ought nevertheless to be measured as an integral part of the earth.

* *Les Espaces Célestes et la Nature Tropicale.*

CHAPTER III.

MEAN PRESSURE OF THE ATMOSPHERE UNDER VARIOUS LATITUDES.—DENSITY OF THE AIR IN THE NORTHERN HEMISPHERE.—DIURNAL OSCILLATIONS OF THE BAROMETRICAL COLUMN.—ANNUAL OSCILLATIONS.—IRREGULAR VARIATIONS.—ISOBAROMETRIC LINES.

The atmosphere is of such mobility that its weight, measured in an exact manner by the column of mercury in the barometer, varies incessantly all over the earth. The various meteoric changes, from cold to heat, from dryness to moisture, augment or diminish the pressure of the air, and in consequence a corresponding oscillation is produced in the mercury contained within the tube of the instrument. Now, any volume of mercury being about 10,500 times heavier than the same volume of air taken at the level of the ocean, we must conclude from this that every movement of the barometric column reveals a change 10,500 times greater in aerial space.

When air is heated either by the direct influence of the sun or by a current of higher temperature, its particles expand, become relatively lighter, ascend into space, and then spread out laterally. The pressure then diminishes, and in consequence the column of mercury in the barometer must fall. The contrary takes place when the air is condensed by cold, and when the aerial masses flow together to fill up the space; the weight of the atmosphere is increased, and the level of the mercury rises in the instrument. This is the reason why the fall of the barometer indicates generally an increase of temperature, while a diminution of heat is marked by the contrary phenomenon. The barometer and thermometer oscillate in inverse ways. It is true that the air can absorb so much the more watery vapour the warmer it is, and in this way the pressure which is diminished on one side by the ascent, and the lateral flow of the aerial fluid, is augmented on the other by the increase of vapour contained in the atmosphere; air becoming colder, on the other hand, loses its capacity of dissolving the watery vapour, and grows lighter in proportion. Thus the phenomena counterbalance each other, and it is not without numerous observations treated with sagacity that we are able to distinguish that which, in slight barometric oscillations, ought to be attributed either to the pressure of the pure air, or to that of the watery vapour. As to the abrupt variations in respect to which we cannot be mistaken, they are sometimes enormous; there are even some which are marked in the column of mercury by a difference of 2 or 3 inches, one-fifteenth of the total height. A tempest in the ocean of the air is the cause of this agitation of the liquid in the instrument.*

The pressure of the atmosphere varies over all the earth, and we cannot yet indicate it with exactitude for the entire globe. It is probable, however, that at the surface of the sea it exceeds on an average, by a slight fraction, the amount of 29·90 inches. Towards the equator the ordinary pressure is only 29·84 inches; but from the 10th degree of latitude in the two hemispheres the pressure increases little by little, and towards the 30th or 35th degree it attains its maximum, 30 or 30·08 inches. From thence, in the direction of the poles, the pressure diminishes; towards the 50th degree it is 29·92 inches, and further north 29·76 inches only. Thus it is at about an equal distance between the pole and the equator that the air exercises on an average its greatest pressure on the barometric column; nevertheless, there being much more watery vapour in the aerial strata of the temperate zone than in those of the polar zone, it is probable that, the air being perfectly dry, its pressure would continually increase from the equator to the poles more or less regularly in proportion to the sinking of the temperature. This is moreover a phenomenon rendered very probable by the rise in the barometer, which is ordinarily produced by the transition from heat to cold. However it may be, the researches of Sir James Ross and Wilkes in the southern seas establish the fact that on an average the barometer is slightly higher in the northern than in the southern hemisphere. We must necessarily conclude from this, that a greater quantity of air is accumulated over that half of the earth where the continents are grouped. Thus, as Sir John Herschel remarks, the current of a river is always rippled above an unequal and stony bed; and in the same way the atmosphere must swell in waves above the continental masses. This best explains the astonishing contrast between the two hemispheres.†

If the normal pressure of the atmospheric strata varies at the level of the ocean under different latitudes, it varies also all over the earth

* See below the section entitled, *Hurricanes*.
† W. Ferrel, *Motions of Fluids and Solids*, p. 39.

according to the hours and seasons; it obeys the rhythm of time as well as that of space. Every day the aerial mass oscillates twice in inverse directions. In the morning towards four o'clock the barometric column presents a first minimum of height, but it rises gradually, and towards ten o'clock in the morning it attains its highest elevation; afterwards the pressure of the air diminishes till towards four o'clock in the evening, the time when the barometer is at its minimum of height. The column of mercury then begins to rise till ten o'clock at night, to sink again for six hours; the periods of the day during which these changes occur are known under the name of "tropical hours." The curve of variations, as we see by the accompanying figure, is much more regular in the equatorial than in the temperate zone.

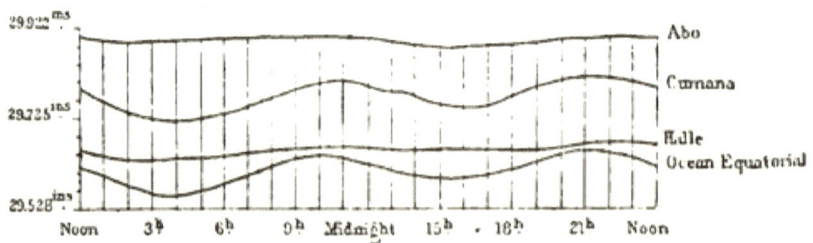

Fig. 90.—Tropical hours of the Equatorial Ocean, of Cumana, of Halle, and Abo.

What is the cause of this double daily oscillation? Many meteorologists formerly recognised in these movements of the barometer regular tides, similar to those in the ocean, and, like them, obeying the combined influences of the sun and moon. But these oscillations always occur, on an average, at the same hours, and do not present at the epoch of syzygies and quadratures phenomena corresponding to those of the ebb and flow.* The researches of Aimé, of Flangergues, and other natural philosophers have, it is true, established the existence of an aerial tide; but the amplitude of this movement is very slight in comparison with that which occurs between the tropical hours. It is, therefore, by the combined influence of the heat of the day, and the pressure of watery vapour, that we must explain, with Dove, the two movements of rising and falling, which take place every day in the column of mercury. Commencing in the cold hours of the morning, the gradual increase of temperature must result in expanding the atmosphere, and making the barometer

* See above the section entitled, *The Tides*.

sink. But while the pressure of the air diminishes, the quantity of watery vapour augments rapidly, and its pressure added to that of the air, produces a sort of temporary wave, after which the barometric column continues to fall, to rise again with the nocturnal cold. If the pressure of the watery vapour disappeared from the atmosphere the barometer would rise regularly in all seasons towards the middle of the night, and would be at its lowest towards the middle of the day. This is shown by the following figure, representing the barometric oscillation of the dry air at the port of Apen-

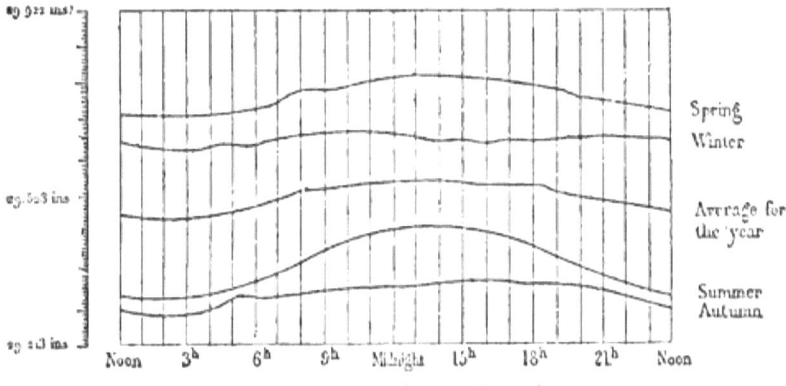

Fig. 100.—Pressure of dry air at Apenrade.

rade on an estuary of the Baltic. In very dry countries, such as Eastern Siberia, the pressure of the watery vapour is too slight to counterbalance the action of temperature, and in consequence only two oscillations occur during the four-and-twenty hours, a fall with the increase of temperature, and a rise with the cold of night.

The diurnal movements of the barometer are much more regular and more easily ascertained in the equatorial regions and near the level of the sea than under high latitudes and in the interior of continents. This is because over the tropical seas the alternations of temperature, evaporation, and precipitation succeed each other, like all other physical phenomena, with greater regularity than in other parts of the globe. Besides, it was in the equatorial seas that the diurnal oscillation was observed for the first time, and it was in these same latitudes that Humboldt was able to ascertain the hours exactly. In the temperate regions these regular movements of the barometric column are in a great measure hidden by the abrupt leaps of the mercury, obeying the constant variations of the atmosphere; it is, therefore, only after a longer or shorter series of days, or even weeks, that meteorologists can, by establishing averages, reveal normal oscillations analogous to those which occur at the equator. In the high mountain regions it is still more difficult to ascertain the regular succession of the barometric waves, for the changes which occur in the lower strata of the air are only felt later, and are variously mixed in the higher strata. Thus the rising of the barometer which takes place towards ten in the morning at Zurich, does not occur on the summit of the Righi till two o'clock in the afternoon, and only at three o'clock on the Faulhorn; often the depression of the barometric column does not even make itself felt in the afternoon on these heights, and each day presents but a single great oscillation.

The annual variations of the pressure of the air present alternations analogous to those of the diurnal variations. In the tropical countries, where the seasons follow one another with great regularity, and in the interior of continents, the air of which contains but a slight quantity of watery vapour, the mercury of the baro-

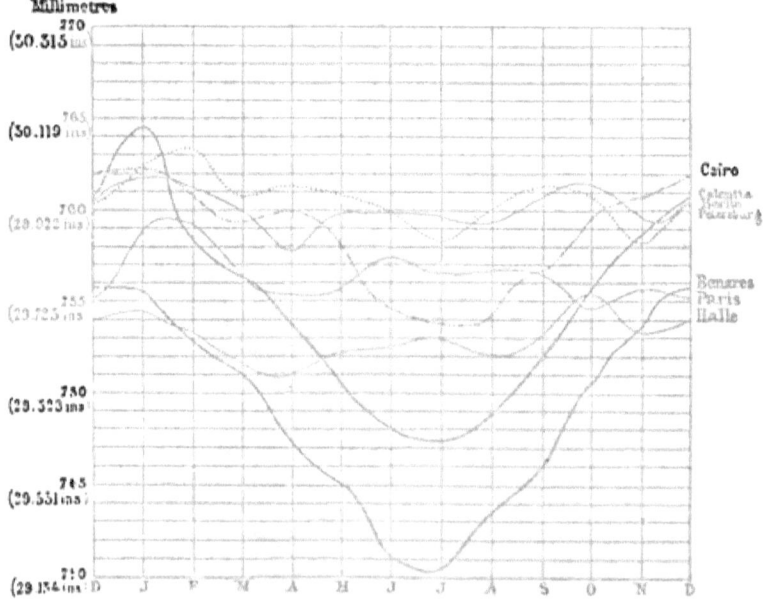

Fig. 101.—Monthly variations in the pressure of the atmosphere at Cairo, Calcutta, Berlin, St. Petersburg, Benares, Paris, and Halle.

meter gradually sinks from winter to summer in inverse proportion to the heat, and re-ascends with the cold from summer to winter.

At Calcutta, at Benares, in Hindostan, as at St. Petersburg, in Prussia, and at Nertschinsk, in Siberia, the maximum of the pressure of the air is perceptible in the month of January, while the minimum occurs in the month of July. The atmosphere is now heavier, now lighter in each hemisphere according to the regular alternations of heat and cold. Thus, as the following figure shows, the annual variation in the pressure of the air occurs in the same manner in all the countries situated on the same side of the equator; but the phenomenon is much more striking in tropical climates than under high latitudes. In the greater part of the countries of the temperate zone, and above all on the shores of the ocean, the pressure of the watery vapour during the summer increases considerably, and thus, counterbalancing the normal effect of dry air, gives to the barometric curve a maximum of summer which corresponds to the diurnal rise of ten o'clock in the morning, or else complicates the series of monthly variations by very numerous irregularities. Each one of these inflexions corresponds to some important phenomenon in local climate, cold or heat, storms, or tranquillity of the air, dryness, or a great quantity of watery vapour. In general it is at the epoch of the equinoxes, when the temperature is nearly equal to the annual mean, that the mean barometric pressure of the year is established.

As to the irregular variations, they are also accomplished according to a certain rhythm in various regions of the globe. At the equator they are almost nothing, but in proportion as we approach either of the poles, the irregularities become more marked, and the leaps produced in the column of mercury by sudden changes of temperature, and by the alternations of winds and storms, succeed each other more frequently. In tropical regions these differences of barometric height are hardly a few fractions of an inch, while in the temperate latitudes they have exceeded 2·1 inches at Milan for a period of 81 years, and 2·6 inches at St Petersburg for a period of 19 years. In order to obtain figures more comparable with each other Kämtz has calculated the monthly amount of the oscillations of the barometer for every station, and in this way has been able to draw up the following table:

Latitude.	Monthly Barometric Amplitude.
0° to 10°	0·1 inches.
10° to 20°	0·17 ,,
20° to 30°	0·32 ,,
30° to 40°	0·53 ,,
40° to 50°	0·81 ,,
50° to 60°	1·03 ,,
60° to 70°	1.2 ,,

Still, we must not expect to find exactly the same amplitude per month at all the points situated at the same distance from the equator.

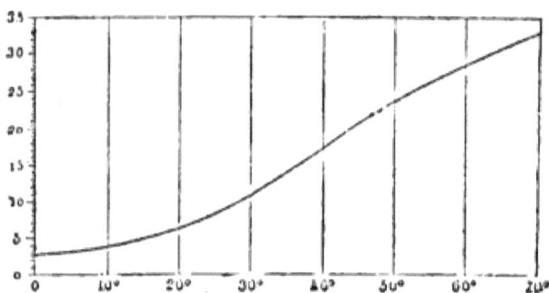

Fig. 102. - Monthly amount of the oscillations of the barometer in the northern hemisphere.

In this respect, on the contrary, we observe great diversities, which we must attribute to the difference of continental forms and of climates. By uniting with each other all the points in which the same monthly variation in the pressure of the air occurs, we obtain a series of lines called *iso-barometric*, which all curve to the north across the Atlantic, and on the whole much resemble the lines called *isothermal*.* These are curves imagined by Kämtz, which indicate the true latitude for the general movements of the atmosphere. In spite of the extreme mobility of air, in spite of the tempests which roll with fury from one point to another of the horizon and disturb for a moment the regularity of atmospherical phenomena, these lines maintain from year to year their mean direction; while indicating the disturbances of the air, they show by their permanence and their regularity how much these commotions depend on the great laws which rule our planet.

* See below the section entitled, *Climates*.

CHAPTER IV.

GENERAL LAW OF THE CIRCULATION OF WINDS.—TRADE-WINDS FROM THE NORTH-EAST AND SOUTH-EAST.—EQUATORIAL CALMS.—OSCILLATION OF THE SYSTEM OF WINDS.

In the continental regions, and principally in those of the temperate zone, it would often be difficult to recognize at first the general law which presides over the movements of the atmosphere, for these various oscillations may be modified by a crowd of local circumstances, such as the direction and height of mountain-chains, the extent of plains, the contours of the shores, the abundance or scarcity of vegetation. Even in one day the winds will sometimes blow successively from all points of space, and among these rapid changes to which the atmospheric currents are subject it is not always possible to ascertain with certainty the normal direction of the mass of air in movement. To understand the laws of the atmosphere in their simplicity we must transport ourselves to the equatorial regions of the ocean, above which the sun describes each day its immense semi-circle in the space of twelve hours, and where all the movements of nature, regulated by the uniform march of the sun, have something of the rhythm of the celestial cycles. It is there that we may seek the first displacement of the atmosphere, which travels as an immense sheet of air all round the globe. We are there present at the birth of the winds. It is there that Eolus would be seated if the gods still lived.

During the days of summer we perceive from afar a vibratory motion of the air over the heated earth, a kind of vaporous trembling, doubtless rendered visible by the incessantly changing mirage of objects lying beyond.* This is because the strata of the atmosphere reposing on the ground have gradually expanded, and rise in spirals through the colder and denser medium which weighs upon them. In the same way the rarefied air of furnaces mounts rapidly towards the upper regions whither its relative lightness carries it.

* See above, p. 227.

A similar movement is produced on a very large scale in the equatorial regions. The great force of the sun's rays making themselves principally felt in these countries of the world, the aerial strata expand under the influence of the heat much more than in other latitudes. They become lighter, and rise rapidly into space, as is shown by the slight pressure of the air on the barometric column.* Thus a void is formed, which the adjacent masses of air hasten to fill, and two horizontal currents go to feed the great vertical current which ascends towards the higher strata of air in the equatorial regions. But these horizontal currents themselves leave spaces behind them towards which new masses of air rush; the atmospheric waves move nearer and nearer through all the zones as far as the polar ice, and from the two ends of the planet, they march towards the equator, where the ascending movement of the over-heated air summons them. Two winds, the one from the north, the other from the south, take their origin from the midst of the ice of the two opposite poles to meet at the equatorial circle.

If the earth were not carried by a movement of rotation around its axis, the atmospheric currents would flow directly towards the equator, without deviating to the right or left from the lines of the meridian. The northern current would flow in a straight line to the south, the southern current would direct itself exactly towards the north, and they would meet in direct opposition in the equatorial regions. But it does not happen thus, because of the rotation of the globe from west to east. The speed of this movement varies for each point of the terrestrial surface, as the diameter of its latitude; whilst it is nothing at the poles, it is 520 miles per hour at the 60th degree of latitude, north or south, at the equator itself it is 1050 miles. The mass of air which flows from the poles towards the tropical zone thus travels successively over latitudes, whose own speed around the axis of the globe is greater than theirs, consequently they are compelled to deviate further and further towards the west in the opposite direction from the general movement of the earth. Instead of being directed perpendicularly towards the equator to form with it an angle of 90 degrees, the aerial currents coming from the poles strike the equinoctial line obliquely at an acute angle. Thus the same planetary phenomenon that causes the deviation of the flow of water† of the oceanic currents,‡ and perhaps even, according to M. Musset, the swelling of the trunks of trees in the direction from east to west, suffices likewise to put in

* See above, p. 228.   † See *The Earth*, the section entitled, *Rivers*.
‡ See above, p. 92.

motion the whole mass of the atmosphere. The rivers of the air reproduce, only in greater proportions, on account of their larger domain, the immense curves of the oceanic currents. The two fluids in movement, winds and marine currents, are superposed in their march around the planet.

In the tropical zone, where the incessant attraction produced by the ascending current determines a constant afflux of masses of air coming from the north and south, the circulatory system of the winds possesses in general a tolerable regularity. In this part of the terrestrial sphere the aerial masses move uniformly, those of the northern hemisphere in the direction from north-east to south-west, and those of the southern hemisphere in the direction from south-east to north-west. Thus two atmospheric currents do not cease to flow obliquely to meet each other. These are the "trade-winds," which the ancients hardly knew, and of which the complete discovery was reserved for the great Spanish and Portuguese navigators. Among all the marvels that they discovered in the tropical regions, none astonished them more than these breezes, blowing invariably from the same point of the horizon. Accustomed to the changing and irregular winds of the European seas, the sailors were almost terrified at the constancy of these winds, which carried them towards the equator, and never flowed back again in the direction of their country. The companions of Columbus saw in it the effect of the craft of the devil, and asked with terror if all this movement of the aerial waves was not directed towards some gulf, situated at the limits of the world. Nevertheless, navigators were soon familiarized with the tranquil latitudes traversed by the trade-winds. The Spanish sailors formerly called the tropical part of the Atlantic Ocean *el golfo de las Damas* (the ladies' sea) because there one could confide the helm of a ship to a young girl without danger. Indeed, according to Varenius, the sailors setting out from Acapulco could fall asleep without paying attention to the rudder, with the certainty of being conducted by the wind across the calm waters of the Pacific, to the shores of the Philippines. Struck with the great advantages which the constancy of the trade-winds present to navigation, the English have given them this name. The old term, *vents alizés*, by which the French sailors designate them, indicates an equal continuous and regular movement.

Still, it must be said, these winds have not such a certain march, that we can count on them, as on the return of the heavenly bodies. The alternation of the seasons and the great atmospheric disturbances make them oscillate from right to left, retard or accelerate them, and sometimes even neutralize them for a time. In the neighbourhood of the coast, the extremes of heat and cold which succeed each other on the continents cause the winds to deviate in their course,* and consequently it is only in the open sea, at a great distance from the coast, that the sails of ships are swollen by a breeze blowing almost constantly from the same point of the horizon; but even then the wind is stronger in the morning and evening than during the heat of the day. In the Atlantic, bordered on each side by continents tolerably regular in form, the trade-winds have the most uniform speed. In the Pacific, the multitude of islands scattered over the surface of the waters modify greatly the normal condition of the winds, and over a very great extent of their natural domain the trade-winds are transformed into monsoons. To the north of the equator the north-east winds only blow in a constant manner, between the Revillagigedo, and the Marianne Islands. As to the southern trade-winds, they are still more restricted; they commence with the group of the Gallapa-gos at 1620 miles from the coast of America, and towards the west, they do not pass the Archipelago of Noukahiva and the Low Islands.

In rushing one against the other, the two opposite winds hold each other in check, and consequently their force is neutralized; it is thus that a circular zone of calms, variable winds, and sudden aerial eddies is formed all round the earth, which, according to the seasons, occupy a breadth of from 155 to 620 miles on the surface of the sea. Nevertheless, we must not suppose that, in this zone of so-called calms, the air is generally tranquil; but the atmosphere is more often in a state of equilibrium there than in any other part of the surface of the globe. According to the *Pilot Charts* of Maury, the mean duration of the calms of the Atlantic between the 5th and 18th degrees of north latitude is to that of the winds in the proportion of 98 to 802, or of 1 to 8. During the period when the calms are most frequently produced, that is to say, in the month of November, and in the space comprised between the 12th and 13th degrees of north latitude, they prevail on an average half as much as winds coming from any point whatever of the horizon.

We can understand that this zone which separates the two trade-winds of the north and south must necessarily be altered according to the seasons by the position of the sun, since it occupies on the circumference of the globe precisely those latitudes where the atmospheric strata are most strongly heated by the solar rays, and where the vertical movement of the expanded air is produced. When the sun,

* See below, Fig. 107.

after the 21st of September, crosses the equatorial line to tend towards the tropic of Cancer, the centre of the trade-winds, and consequently of the band of calms, moves at the same time towards the north; on the contrary, when the sun returns to the tropic of Capricorn, the most heated zone of air is gradually brought back to the south with the whole circulatory system of the trade-winds. At the end of March the northern limit of the equatorial calms of the Atlantic is found, on an average, towards the 2nd degree of north latitude, while at the end of September this same limit attains ordinarily the 13th or 14th degree.* As to the southern limit, it oscillates in the same ocean from 1 to 4 degrees of north latitude.† In the equatorial regions of the Pacific, the zone of calms is similarly displaced from month to month, following the march of the sun, and its breadth varies from 135 miles in the month of February to more than 840 miles in the month of August.‡ In this respect the analogy is almost complete in the two great oceans. In consequence of this annual periodicity, the whole aerial system incessantly oscillates according to the relative position of the sun, and it is for this reason that, in the northern hemisphere, the north winds, being violently attracted towards the south, are much stronger in winter. Besides these, there probably exist monthly and semi-monthly oscillations resulting from the declination of the moon.§

The central part of the zone of calms, which may be considered as the meteorological equator of the world, does not correspond with the equator properly so-called. On the earth, as in the higher organisms, the principal seat of life is placed out of the geometrical centre. The complete system of the winds inclines towards the northern hemisphere, and it is to the north of the line that the girdle of the equatorial calms is at all seasons developed. This phenomenon, which might seem at first sight strange, results from the grouping of the greater part of the continental lands in the northern hemisphere, and from the difference of temperature, which must be, at least for one part of the world, the result of this unequal distribution of solid and liquid. It is also in the northern hemisphere that we find the desert of Sahara, the true geographical south of the earth, that immense extent where wooded tracts are relatively few, and where the reflection from the burning sands and rocks vaporize the clouds

* D'Après;—Dove, *La Loi des Tempêtes* (traduction Légras), p. 17.
† Horsburgh, *East India Directory*, vol. i. p. 25.
‡ Kerhallet, *Considérations Générales sur l'Océan Pacifique.*
§ Keller, Rennell.

which the atmospheric currents bring. The Sahara, and in a less degree all the tropical countries of the northern hemisphere, act as a great centre, towards which the aerial masses flow. It results from the tables drawn up by Dove, that the mean temperature of the year is more elevated (92° Fahr.) towards the 10th degree north latitude, than it is at the equator itself (91·5° Fahr), while the mean balance is stronger towards the 20th degree of latitude (94° Fahr.) than in any other region of the world.* The high temperature of the continents thus forces the southern system of winds to encroach upon the northern system.

* Dove, *Vorbereitung der Wärme auf der Oberfläche der Erde.*

CHAPTER V.

COUNTER TRADE-WINDS OR RETURNING WINDS.

THE aerial masses brought by the two trade-winds cannot be incessantly accumulated in the region of equatorial calms; they expand, rise to several miles of height, then, after having been mixed, and even partially crossed, they divide anew into two great returning currents which flow in an opposite direction in the upper regions of the atmosphere. Thus, as the natural philosopher Halley, who was the first to give a theory of the trade-winds, affirmed nearly two centuries ago, it would be absolutely impossible, if these two atmospheric counter-currents did not exist, that the equilibrium of the air could be established on the surface of the globe; that which the breath of the trade-winds bring to the equator must necessarily be carried back by other winds towards the poles. The movement of the graceful clouds, so light that we see them from below floating in the heights of the air in the opposite direction to the trade-winds, is an incontestable proof of the existence of these higher returning currents.* Besides, two great volcanic explosions, often mentioned by savants, have also furnished striking testimonies which confirm Halley's theory in an indubitable manner. On the 1st of May, 1812, when the north-east trade-wind was in all its force, enormous quantities of ashes obscured the atmosphere above the island of Barbadoes, and covered the ground with a thick layer. From whence came these clouds of dust? One would have supposed that they came from the volcanos of the Azores, which were to the north-east, nevertheless, they were cast up by the crater of Morne Garou, situated in the island of St Vincent, at 125 miles to the west. It is therefore certain that the débris had been hurled by the force of the eruption above the moving sheet of the trade-winds, into an aerial river, proceeding in the contrary direction. In the same way, at the time of the terrible eruption of the volcano of Coseguina, in Central America, ashes were carried by the returning trade-winds to the shores of

* See *The Earth*, the section entitled, *Volcanos*.

Jamaica, which is no less than 800 miles to the north-east of the crater whence they were thrown.

Fig. 103.—Cloud of cinders from Morne Garou.

On the coasts of Africa, and along the Mediterranean, grains of dust, almost imperceptible singly, give another very remarkable proof of the existence of a great returning current in the high regions of the atmosphere. Sometimes a shower of yellow or red dust, resembling powdered brick, falls from the sky. Ships which were in the neighbourhood of Cape Verde, on the coasts of Morocco, or in the waters of the Mediterranean, have had their deck and sails completely sprinkled with these fine particles. Humboldt, who had the opportunity to observe this rain, believed that it was composed of silicious dust raised by eddies of wind on the coasts of the Sahara, while the sailors who witnessed this phenomenon saw in it only a shower of sulphur. But Ehrenberg, with the aid of his microscope, revealed the nature of this dust, which is nothing else, at least in the Atlantic and the Mediterranean, than the silicious skeletons of animalculæ coming from the llanos of South America. It is thus certain that these myriads of organisms, raised to a height in the air by the ascending current of the equator, have met above the trade-winds with a returning current, which has caused them to cross the immense basin of the Atlantic, and to reach the coasts of Africa, or even of Europe, as far as the

SHOWERS O

Greenland
Ice
NORTH AMERICA
Newfoundland
1719
ATLANTIC
The Azores
OCEAN
1861
Cape de Verdes
1838
1839
SOUTH AMERICA
Cayenne
1802

basin of the Rhône. The aerial currrents have thus become visible, by means of these clouds of infusoria.*

In the equatorial zone the counter-current of the trade-winds can only commence at a height of from four to five miles above the level of the sea, for the highest summits of the Cordilleras remain entirely bathed in the lower current. The most southern mountain of the Atlantic basin, whence the returning wind has been observed, is the peak of Teyde, in the island of Teneriffe. There the masses of air

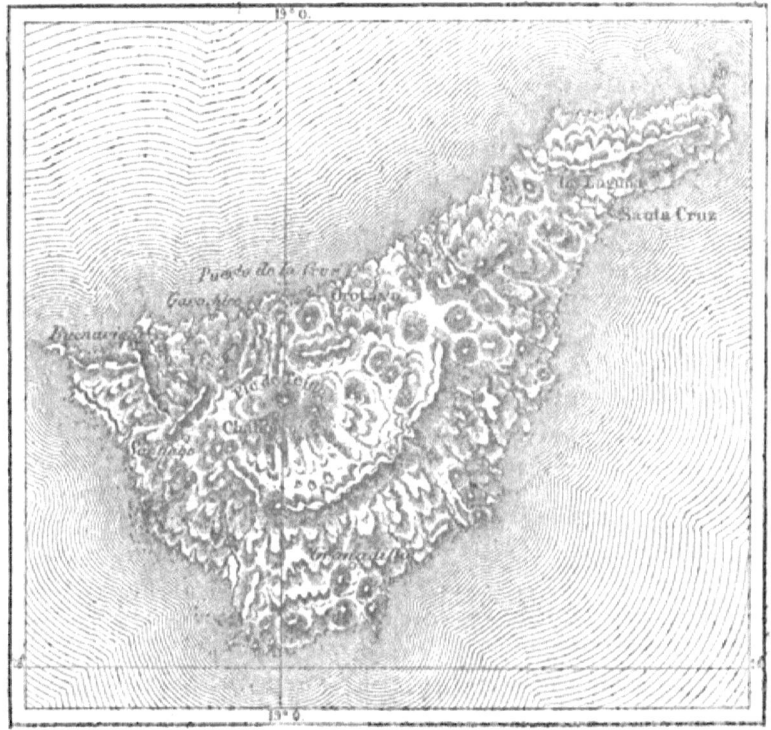

Fig. 104.—Island of Teneriffe.

flowing back from the equatorial zone, are already sufficiently low to surround the terminal point of the volcano (only 12,060 feet high) at all seasons. In winter, when the whole circulatory system of the atmosphere has descended towards the south, following the course of the sun, the returning current descends from the higher strata of the

* Maury, *Physical Geography of the Sea.*

air, and strikes the surface of the water near the coasts of Portugal, then turns back again, and makes itself successively felt at Madeira, and on the middle and lower slopes of the peak of Teneriffe.* According to the astronomer Piazzi Smyth, it is at an average of 9000 feet of vertical height that the plane of separation between the two aerial rivers flowing in opposite directions is found. At the summit of the mountain the air is carried rapidly from south-west to north-east, while on the low parts of the island the trade-wind always blows with its habitual regularity.† The zone of clouds that it unrolls in an immense veil above the sea and the shores, does not extend into that part of the heavens comprised between the two winds blowing different ways, but, on the contrary, it is found at a tolerable depth in the trade-winds. Between the upper and lower currents the air is calm and free from clouds. During the summer season, travellers who climb the sides of the peak of Teneriffe, may confidently expect to find an unchanging sky directly after having passed the zone of clouds, from 900 to 1200 feet in thickness, which is spread like a second sea above the ocean.‡ At the change of the seasons, when the two opposing winds strive for victory on the slopes of the mountain, a few days are sometimes sufficient to bring about a change of 3000 feet in the height of the intermediary zone. A battle between the two currents takes place in the sky; soon the trade-wind mounts to the upper slopes of the peak; now it is vanquished, chased from the heights of the atmosphere, and driven with all its system of clouds towards the lower regions. It is principally above the pass of Laguna, between Santa-Cruz and Orotava, that the combat takes place, and in consequence, this district of the island is frequently inundated with rains. Piazzi Smyth has recounted these grand aerial contests at great length in his work on Teneriffe.§

In the Pacific Ocean, analogous phenomena to those which occur in the Atlantic have been observed. Goodrich has ascertained that the normal current of the trade and returning winds make themselves felt at the same time, the one on the shores of the Sandwich Islands and the lower slopes of all the mountains of the Archipelago, the other on the summit of the volcano of Mauna-Loa.

The direction of the upper counter-current is, like that of the

* Humboldt; Leopold von Buch, *Description des Canaries*.
† *Philosophical Transactions*, 1859.
‡ See below, the section entitled, *Clouds and Rains*.
§ *Teneriffe*, pp. 178, 174, 432, &c.

trade-winds, determined by the rotatory movement of the earth. At its return from the equator, each particle of air in movement turns towards the east, instead of deviating to the west, as in its voyage from the polar to the torrid zone. After having sojourned in the equatorial regions it traverses successively countries whose speed around the axis of the earth is less than its own. In proportion as it retreats from the zone of calms, it thus finds itself in advance of all the subjacent points of the planet, and changes into a wind from the south-west. Below it, glides the north-east trade-wind, generally in an opposite direction; but in consequence of the friction of the aerial particles, a stratum of calm air is formed between the two atmospheric currents, where all the meteoric phenomena due to the contact of the two masses, unlike in heat, moisture, and electric tension, are manifested. According to Dove, the counter trade-wind would bend more and more to the east, because of the increasing curve of

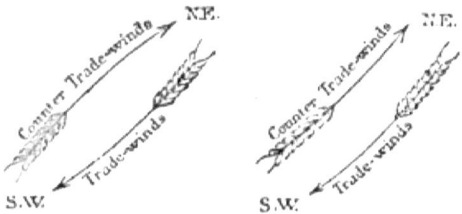

Fig. 105.—Theory of Dove. Theory of Mühry.

the earth in the direction of the pole. According to Mühry, on the contrary, the direction of this wind would be exactly parallel to that of the lower current, and would curve gradually towards the north, in consequence of the attraction exercised in the polar regions by the wind which descends towards the equator. This last theory appears the most probable; but it is for direct observation to decide in a definite manner.*

One might believe at first that the upper counter-current flows towards the pole, maintaining itself in the high regions of the atmosphere, and that the polar wind, on its side, more compact in its particles, because of the cold, always glides over the surface of the globe. It is only rarely that it is so. In a somewhat undecided region which, for the North Atlantic, oscillates alternately, according to the seasons, from the 21st to the 25th degree of latitude, the returning wind commences to descend from the heights of the sky to the surface of the sea, and strikes against the aerial masses which flow from

* *Mittheilungen von Petermann*, ix. 1866.

the poles towards the burning latitudes of the equator. The zone where this shock of the winds occurs is considered as the outer limit

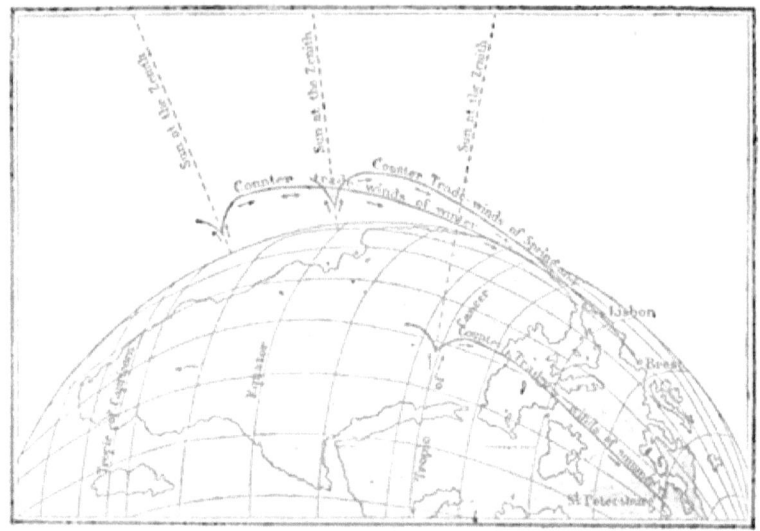

Fig. 106.—Variations in the Trade-Winds.

of the trade-winds; but it is incorrect to give it the name of the zone of tropical calms, for if the complete equilibrium of the atmosphere is more frequent there than in the regions bordering on the north and south, yet, nevertheless, the calms hardly last more than a day in the space of two or three weeks.* During the summer of the northern hemisphere, when the sun is at the zenith of the tropic of Cancer, the counter trade-winds may make themselves felt with tolerable regularity, as far as the latitude of the north of Germany, or of St Petersburg. In autumn and in winter the domain of these returning currents is unceasingly restricted towards the north, and increases to the south. Brest and then Lisbon are its extreme limits in the northern hemisphere, till the sun resumes its march to the north.

Why does the upper current thus descend from the heights of the atmosphere during the greater part of the year? Doubtless, because it carries with it enormous quantities of watery vapour, which renders it heavier than the cold dry air from the poles. Owing to its temperature, it first rises higher than the Cordilleras, then, being gradually cooled, it sinks under the weight of moisture that satu-

* Lartigue, Maury, *Pilot Charts.*

rates it, and when it finally enters the temperate zone, it falls to the surface of the earth with clouds and rains, and strives for supremacy with the polar current. The difference of specific cold between the opposed masses of air must be very small, since by turns each gains the advantage. Often the current coming from the torrid zone, recognizable from below by its trains of *cirri*, cannot reach the surface of the ground, and maintains itself as far as the pole in the upper strata of the atmosphere, whilst the wind that blows from the frigid zone forms a continuous current over the earth, from the pole to the equator. Still, we must consider the south-west wind as the prevailing wind of the northern temperate zone, for it makes itself felt there much more frequently than the contrary current, the proportion of the former being nearly double between the 50th and 55th degree of latitude.* We know that sailing vessels formerly required 46 days on an average for the voyage from Europe to the United States, whilst the return, facilitated likewise by the Gulf-stream,† was accomplished in 23 days. The winds from the south-west and west, which are nothing else than the counter-current of the trade-winds, blow with such regularity in these parts, that one might give the names of "ascending voyages" to the passages from Europe to America, and "descending voyages" to those in the opposite direction. Corresponding phenomena occur in the southern hemisphere; there it is the north-west winds which blow most frequently beyond the southern limits of the trade-winds.

Thus the two permanent winds which are drawn towards the equator by the expansion of the warm air have each their proper domain, limited in one direction by the calms of the equinoctial line, in the other by the irregular winds of the temperate zone. Still these limits oscillate incessantly from month to month and from season to season, and one cannot indicate them in a precise manner. On a general map of the trade winds, it is sufficient therefore to trace the extreme frontiers of these currents for winter and summer.‡ On an average, the space in the Atlantic over which the north-east wind blows, embraces from 18 to 20 degrees of latitude, or from 1245 to 1275 miles; in the South Pacific the domain of the south-east trade wind would not be less than 30 degrees,§ or 2015 miles.

Between the atmospheric and oceanic currents the analogy is evident. The maritime river which is founded at the junction of

---

* Maury, *Pilot Charts*.   † See above, p. 76.
‡ See above, Fig. 106.
§ Kerballet, *Considérations Générales sur l'Océan Pacifique*.

masses of water coming from the two polar seas, corresponds to the equinoctial zone, where the trade-winds of the north-east and south-east meet. Obliged to extend laterally while maintaining themselves under the common level of the maritime reservoir which contains them, the tepid waters of the equatorial current flow afterwards towards the north-east parallel to the counter-current of the trade-winds which has risen into the heights of the atmosphere. Under the influence of the same causes, the two oceans of air and water move in the same direction, and their movements are subject to the same oscillations to the north or south during each alternating period of the seasons. In summer, when the Gulf-stream is prolonged far into the northern sea, the double system of the north trade-winds and counter-winds advance several degrees into the temperate zone, while during the winter it flows back again towards the tropic of Cancer, followed by the Gulf-stream, which bends gradually towards the south. The resemblance would be complete if the water were an elastic and compressible fluid like air, and were not enclosed in a basin whose borders it could not pass. The difference of the means explains the difference of the currents which the heat of the sun and the terrestrial rotation produce in the ocean and in the atmosphere.

CHAPTER VI.

THE TRADE-WINDS OF THE CONTINENTS.—THE MONSOONS.—ETESIAN WINDS.

THE trade-winds, as we said, have not the same regularity on the continents as over the seas. On the surface of the ocean the masses of moving air are not arrested by any obstacle; they are propagated freely towards the equatorial zone, and can scarcely be turned from their route by the attraction of any marine centre of heat, as the temperature of the water only increases or diminishes very slowly, and the oscillations of the thermometer from day to night do not attain 36 degrees Fahr. In the midst of the large islands and the continents it is no longer so. Their mountain-chains oppose the course of the winds, and cause them to change their direction; forests, prairies, sheets of inland waters, plateaux with long slopes, hilly countries, large plains, and the innumerable variations of topographical relief, are variously heated by the sun, and by this very circumstance turn aside or repel the wind which blows from the neighbouring seas. In the higher regions the current can, it is true, continue its normal movement above the plateaux and the mountains; but below the uneven surface of the country is traversed by irregular winds. Here the band of equatorial calms is completely obliterated, there it is enlarged in an abnormal manner; the winds are deflected variously on one side or the other, and are directed towards that country whose air is most expanded by the rays of the sun. Nevertheless, it must be said that only a very insufficient number of meteorological observations have as yet been made in the greater part of tropical countries.

Still we cannot doubt but that the trade-winds blow over vast continental tracts, as well as over the surface of the seas. In fact, the want of rain and the almost complete absence of vegetation in all that part of Africa known by the name of the desert of Sahara, prove in an indubitable manner the existence of a regular wind from the north-east. After having passed the high plateau of Asia, and having discharged itself of the greater part of its watery vapour, this atmospheric current traverses obliquely the whole of Africa from the banks of the

Nile to those of the Niger. On this enormous track of nearly 3000 miles it only lets rain fall on some mountain-summits, such as the Djebel-Hoggar, and scarcely casts a single cloud on the unchanging azure of the sky. On the western coast of the Sahara, the burning wind called the *Harmattan* is nothing else than the north-east trade-wind more or less turned from its course because of the neighbourhood of the sea. Towards the 17th degree of north latitude, on the southern frontiers of Soudan, clouds are at last formed in space, abundant rains penetrate the soil, and the aridity of the desert gives place to a fine vegetation; this is, because the domain of the permanent winds terminates there, to be replaced by the zone of equatorial calms with an ascending current loaded with aqueous vapours. In the southern part of Africa, the trade-winds of the south-east make themselves regularly felt, and according to the testimony of Livingstone, traverse the entire continent from the mouths of the Zambesi to the coast of Angola. On the other side of the great strait of the Atlantic, the tropical regions of South America are likewise refreshed by the constant breath of moist winds from the south-east. Brazil, Paraguay, a great part of the Argentine republic, Bolivia, Peru, Guiana, and Columbia, are comprised in this great meteorological region. The trade-wind, turned back under the equator in an east and west direction, ascends with a uniform force the valley of the Amazons, penetrates into the gorges of the Andes, and even crosses by all the defiles the high barrier of mountains; but sheltered by this enormous rampart the shores of the Pacific are not subjected to the influence of the east wind. The vessels which sail in the open sea have to traverse from 125 to 625 miles, according to the latitudes, before a gust of the trade-wind, descending from the summit of the Andes, comes to swell their sails and drive them to the coast of Australia.

Even in those parts of the world where the tropical winds cease to be permanent, the oscillations and deflections of the atmospheric current present in general a periodical character and occur regularly, according to the course of the seasons. Among the regular return winds we may cite principally the "monsoons" of India and Arabia. The Arabian name for these meteoric phenomena, *maussim*, or *moussim*, signifies change, season; it is because, in fact, they divide the year in the most exact manner into two totally distinct periods. During the great heats of summer, the arid plateaux of Central Asia, and even the plains of Hindostan, much more heated than the sea, act like a great respiring pump; the air which rests above this part of the Asiatic continent expands, and in consequence new aerial masses flow without cessation from the Indian Ocean to the countries on the north. According to Dove, the trade-wind of the south-east, carried away by this general displacement of the air, would itself cross the equator, enter into the northern hemisphere, and transform itself gradually into a monsoon from the south-west, because of the great speed it acquires at the equator. Still it is not probable that it is so, for the monsoon has not the same vertical height as the trade-winds, and its direction is not uniformly from south-west to north-east, as on the coasts of Malabar; in the valley of Scinde and in that of the Irawaddy it is directly south; at the extremity of the Gulf of Bengal, at Siam, at the eastern angle of the Asiatic continent, its direction is from south-east to north-west, perpendicular to the coasts which attract the wind.*

Saturated with the moisture which has evaporated from the great cauldron of the Indian Ocean, the monsoon inundates the coasts of Malabar with torrents of rain, deluges the shores of the transgangetic peninsula, and then strikes against the high mountains of the Himalaya, and other chains, which border the plateaux of Central Asia on the south, but it does not cross this barrier. By its clouds charged with rain, which are rent by the escarpments of the inferior peaks, we see clearly that the sea wind does not pass the altitude of 4950 to 8250 feet, and that above it another aerial stratum is moving in the heights. The movement which carries along this elevated stratum is the same as that of the monsoon from the south-west; but we recognize by its long trains of *cirri*, from 16,500 to 25,500 feet high, that great returning current, or counter trade-wind, that blows at the same elevation above the Atlantic in the neighbourhood of the Canaries.

When the sun, in its course over the ecliptic, returns towards the tropic of Capricorn, the centre of attraction is at the same time displaced in a southerly direction. The monsoon of the south-west ceases to tend towards the great peninsulas of Asia, the regular wind from the north-east recommences to blow, and the currents of attraction in the southern hemisphere turn back towards the islands of Sunda and Australia. Owing to this regular alternation, which was a surprise to the ancient Greek navigator Hippalos, the mariners of the Indian Ocean may count beforehand on a favourable wind which by turns will drive their ship before it for the two passages, going and returning; and they have not to dread those prolonged calms which are the bane of sailing vessels in the equatorial zone of the Atlantic and the

* Mühry, *Zeitschrift für Meteorologie von Carl Jelinek*, No. 21, 1867.

South Sea. The circulatory system does not in any place pass beyond the lower strata of the aerial ocean, and we may easily perceive above the islands of Sunda and Australia, as well as over the sides of the Himalayas, the constant progress of the clouds which are brought by the regular trade-winds. A volcano of Java, observed by Junghuhn, affords a remarkable example of this. From its summit, about 9900 feet high, a column of vapour escapes all the year round, which bends gracefully in space, and directs itself towards the west, or northwest, in a long whitish cloud, and it is in precisely the opposite direction that the monsoon blows during six months of the year, on the slopes as well as at the foot of the mountains.

The monsoons of the East Indies are not the only winds which break the uniformity of the trade-winds. In all those parts of the tropical zone where the shores of the continents are disposed parallel to the equator, the winds alternate regularly, in consequence of the greater rarefaction of the air which occurs now on the earth, now on the sea, according to the position of the sun. Thus, during the greater part of the year, the African coasts, which stretch from the Bight of Benin to Cape Palmas, attract the monsoons of the Gulf of Guinea. These masses of air changing their direction, turn back to blow in a north-easterly direction, and rush rapidly towards the great furnace of the Sahara, where the overheated atmosphere is usually more expanded than in any other country of the world. Towards the month of January, when the Sahara itself has become colder than the equatorial seas and the banks of the Congo, the trade-wind of the north-east re-assumes the supremacy, and traverses the whole of Northern Africa obliquely to the south, towards the coasts of Southern Guinea. Very violent at first, it soon becomes weaker, and hardly lasts but two or three weeks, when it again gives place to the marine monsoon. During its short prevalence the current coming from the desert does not cease to bring with it a white dust having the appearance of a thick fog. It is the sand of the Sahara, which in the regions situated immediately to the north of Guinea is almost white, while further the dust raised from the ground by the *Harmattan* is nearly red.*

On the coasts of Chili, on those of California, in the islands of the Pacific, around the Gulf of Mexico and the sea of the Antilles, analogous phenomena occur. In summer the valley of the Mississippi and the plateaux of Texas are traversed by real monsoons, which distribute the rain over that part of the continent, and then are in turn

* Borghero, *Bulletin de la Société de Géographie*, July, 1866.

replaced by those dangerous winds from the north or north-east (*nortes*), which are themselves trade-winds, more or less turned aside

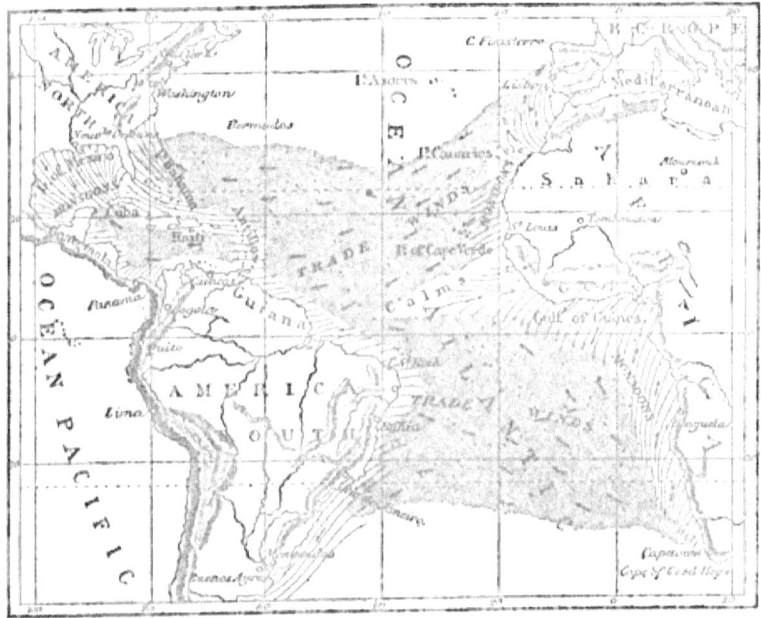

Fig. 107.—Trade-Winds and Monsoons of the Atlantic.

from their route. On their side, the western shores of Mexico present a similar alternation of winds, those from the south-west coming in summer, and those from the north-east in winter. On coasts parallel or merely oblique to the path of the trade-winds, a part of these winds is not brought back, as in the West Indies or at Guinea, but it is more or less attracted by the centre of heat, which lies out of its regular course. It is thus that on the coast of Morocco, and near the Archipelago of the Canary Islands, the wind from the north-east is subject to a considerable deviation towards the African continent, and is sometimes transformed into a wind from the north. In the same way the plateaux of New Granada and the *llanos* of Venezuela turn aside the normal current which penetrates into the sea of the Antilles, and oblige it to blow perpendicularly to the coast. Thus a periodical breeze (*los brisotes*) is produced, which may be considered as an intermediate wind between the monsoon and the trade-wind properly so-called.

The winds of the Eastern Mediterranean, to which the ancients gave the name of *Etesian* winds (from *étos*, year), are also nothing else than monsoons. These are atmospheric currents drawn from the north towards the continent of Africa, by the powerful centres of attraction formed by the sands of Egypt and the Sahara. During nearly the whole summer the aerial masses which repose above Southern Europe are thus carried away to the coasts of Africa, and even in temperate countries with variable winds, like Italy, Provence, and Spain, it is affirmed that the predominating currents are those from the north. Owing to this general movement of the air, the passage from Europe to Africa is accomplished on an average more rapidly than the returning voyage; for the sailing vessels, which traverse the Mediterranean between France and Algiers, the passage from the north to the south is about a quarter less than the route in the opposite direction. All the northern part of the Balearic Islands, and especially of Minorca, is laid waste by a wind from the north, which stunts vegetation, and causes all the trees to lean in a southerly direction.*

* Marié-Davy, *Les Mouvements de l'Atmosphère et des Mers.*

CHAPTER VII.

LAND AND SEA BREEZES.—WINDS FROM THE MOUNTAINS.—SOLAR BREEZES.—LOCAL WINDS.—THE SIMOON, SCIROCCO, FŒHN, TEMPESTS, AND MISTRAL.

BESIDE the lateral deviations which occur twice in the year, the trade-winds are subject along the coasts to rapid daily deviations. The whole outline of the continents is bordered, so to speak, with a fringe of breezes produced by the difference of temperature between the land and the water. During the day the countries of the coast-line are heated much more rapidly than the surface of the ocean. Towards ten o'clock in the morning, after a shorter or longer period of calm, a rupture of the equilibrium occurs between the aerial masses, and the fresher atmosphere reposing above the waters tends towards the land, there to replace the expended air which rises into the higher regions. Little by little this movement of translation, which at first only made itself felt in the neighbourhood of the coast, communicates itself to all the surrounding atmospheric strata, and soon the breeze moving nearer and nearer through the ocean of air, occupies a tolerably large space above the sea and the continent, which it unites as an iron plate unites the two branches of a magnet. During the night the ground loses by radiation a great part of the heat that it had received, while the sea preserves pretty nearly the temperature of the day. The equilibrium is again disturbed, but it is now in the direction of the sea; the breeze is brought back, and blows in the opposite way. It is thus that in the space of 24 hours the breeze oscillates from land to sea, and from sea to land, by a motion of ebb and flow, analogous to that of the tides. In the countries of La Plata these alternate breezes from land and sea present such a regularity that they have received the name of *virazones* (gyrations). Around Otaheite they also succeed each other with such punctuality that a vessel could for several consecutive nights make the tour of the island, and always with the wind behind it.

These breezes, which one might also call daily monsoons, co-exist with the movement of the trade-winds, and are in consequence carried along in the general circuit. Instead of being at right angles to the coast, they more often form with it an acute angle; they blow cross-ways, as Captain Dampier said. Nevertheless, it is not only in the domain of the trade-winds or along the borders of the ocean that the littoral breezes occur; they blow everywhere where a considerable difference of temperature exists between the land and the water, wherever the fresh air of the sea or of a lake goes to fill the vacuum left on the coast-line by an ascending current of warm air. A remarkable example of it is seen in the narrow Adriatic Sea. There, during each fine day, the breeze rises in the centre of the gulf, and takes its direction at the same time in two contrary ways; on one side towards the shores of Italy, on the other towards the islands and mountains of Istria and Dalmatia. During the night the coasts that surround the waters of the Adriatic send back to the sea, as to a common centre, the fresh air which they have received; to the divergent currents of the day succeeds a wave of convergent breezes.

In the same way the mountains have their own system of breezes alternating with a regularity similar to that of the land and sea breeze on the coasts of the ocean. In the day, especially in summer, when the summits of the mountains are exposed to all the intensity of the solar rays, and receive a considerable quantity of heat which causes their temperature to approach that of the valleys, the air reposing on the summits expands and rises. At the same time the air of the plains which lie at the foot of the mountains is itself expanded in greater proportions, so that an ascending current is produced from the base to the summit of the peaks, in all the valleys, and over all the escarpments. The atmospheric strata of the plain move in the direction of the heights with all the more impetuosity the more strongly heated the summits have been by the sun. In certain valleys, especially those of the Stura, and other Alpine rivers, which water the plains of Piedmont, the ascending wind has such force that the greater part of the trees are uniformly inclined towards the mountains. Pollen, remains of plants, insects and butterflies, are carried away by the current of air, and by their débris soil the whiteness of the snow. In the night, phenomena of an opposite kind are produced, but with less intensity; the high mountains whose summits rise far into the sky, lose their heat by nocturnal radiation more rapidly than the valleys, the sheets of air which surround them are chilled and descend again, in part towards the plains from which they had ascended a few hours before. Thus an exchange between the two zones is established, an ebb and flow, a rising and falling atmospheric tide, regulated in its intensity by the variations of the temperature; and here we see again, as in the coast breezes, the rotatory movement pointed out by Dove.

As an example of these breezes, called in the French Alps *pontias, rebats, aloups du rent*, we may cite the three aerial currents which flow incessantly in the valleys of Savoy, unless the local system of atmospheric currents be modified by tempests. These three streams of air are those of Faucigny, Tarentaise, and Maurienne. The first traverses the valley of the Arve from Geneva to Mont Blanc; the second moves in the valleys of the Isère, and its tributary, the Doron; the third alternately ascends and descends the valley of the Arc towards Mont Cenis and the pass of Iseran. Ordinarily, the ascending wind commences towards ten o'clock in the morning in the valleys of Savoy, and the descending current flows back again towards the plains at nine o'clock in the evening. In certain places, it is called *matinière*, because it makes itself felt, most of all, before the rising of the sun. M. Fournet, who has for a long time studied these phenomena of atmospheric tides, has ascertained that the passage, from the ebb to the flow, is especially rapid in the narrow defiles, while in the large basins the alternation is produced after a series of aerial oscillations, and gusts of wind in the opposite direction. Each valley owes a special atmospheric condition to its particular form; in one, the successive breezes are slow and undecided in their pace; in another, they alternate abruptly, and with violence, producing in the space of a few hours variations of temperature of 35, 45, and even 55 degrees. In general, the breezes are regular in the regular valleys, and only present remarkable peculiarities at their issuing into the plain, or else at the confluence of two gorges. Among these winds with peculiar motions, a breeze of the Rhenish basin may be mentioned, known under the name of the *Wisper-wind*. Emerging above Lorch, from the narrow valley of the Wisper, which is filled with woods, and so situated as to be subject in its different parts to all the extremes of temperature, this breeze generally blows till eight, nine, or ten o'clock in the morning, then crosses the Rhine, strikes against the rocks of the left bank, and divides into two currents, one of which re-ascends to the south towards Bingen, increasing itself on the way by several small tributary winds; while the other, which is weaker, descends to the north towards Bacharach.

Even in the plains and countries but slightly varied in surface, daily breezes may succeed each other regularly, because of local differences of temperature produced by the progress of the sun. In the morning, as soon as the sun rises, the temperature, which had fallen to its lowest because of the nocturnal evaporation, increases rapidly, the air expands and spreads towards the colder spaces which extend on the side of the west; a little wind from the east results from this, which becomes gradually a wind from the south-east, in proportion as the sun mounts above the horizon. At noon the expanded air spreads in the direction of the north; and finally, towards the evening, it is on the eastern side, where the aerial strata are chilled, that the surplus air, still heated by the solar rays, directs itself. Thus, when the atmosphere is not agitated by a general wind, a breeze turning regularly round the horizon in the same direction as the sun must be produced. In the northern hemisphere this movement of gyration is accomplished from east to west by the south; in the opposite hemisphere it is by the north that this diurnal breeze effects its gradual revolution from east to west. In mountains the phenomenon is more complex, on account of the ascending and descending breezes, which are intermixed with the gyrating ones. It is remarked, however, that the greater part of the local winds, determined by the difference of temperature, tend towards the west in the morning, then turn gradually in an opposite direction, and blow towards the east when the sun is sinking. These are the *solaures* (*solis aura*) or solar winds of the Department of the Drôme.*

As to the local winds, which characterize certain regions, they originate in the unequal distribution of heat. Such are the *chamsin* of Egypt, the *pampero* of the Argentine Republic; such, above all, is that aerial current to which the name of Simoon, or "poisonous," is given, in the Sahara. As soon as this wind commences to blow the panting traveller can scarcely breathe; the air is burning and dried up, as if emerging from the mouth of an oven; the heat, increased by the radiation of innumerable grains of sand which float in the atmosphere, rises rapidly to 113, 122, and even 133 degrees Fahr.; the sun is veiled, and every object assumes a violet or dark-red hue, while space is filled with dust. In order not to be smothered by this irrespirable air, travellers must envelop their faces in their garments, and the camels bury their necks in the sand. But the simoon is not always accompanied by clouds of dust. Palgrave, who endured a violent simoon in the desert of Arabia, saw not a single cloud of sand or vapour in the sky,

* Fournet, *Hydrologie du Rhône.*

and could not explain the sudden gloom which had invaded the atmosphere.*

In Sicily and in the south of Italy a warm wind occasionally blows from the south, which is considered as a sort of simoon, and is saturated with moisture in passing over the Mediterranean; this is the scirocco. Usually it is not very rapid, and its gusts are interrupted by stifling calms; the surface of the water is hardly agitated, a mist of vapours broods on the horizon, and the sun hides itself behind a veil of whitish clouds. Under the enervating influence of this wind from the south, all exertion becomes painful; but still the terrible phenomena, which occur during the simoon, have never to be dreaded.

In the Alps of Switzerland the south wind is known under the name of *føhn*, a word derived from *favonius*, the southern wind of the Romans. What is, then, the origin of this current? Has it originated in the Sahara, as Messrs. Desor, Martius, and Escher von der Linth believe, and has its burning breath first served to melt the ancient glaciers of the Alps? Or is it simply a counter trade-wind descended from the heights of the atmosphere, and does it come from the Atlantic and the Caribbean Sea, as Dove asserts? Would not the moisture which it brings tend to enlarge the vast rivers of ice? The latter seems probable; but however it may be, the føhn frequently changes its course, and whether or not it is a continuation of the Mediterranean scirocco, the inequalities in the relief of the mountains modify its character singularly. In rising over the slopes, the air expands more and more in consequence of the lesser atmospheric pressure, and it loses a great quantity of heat; from the warm wind, which it was at the foot of the mountain, the føhn becomes a cold current. The ridge once crossed, the aerial mass, which descends again towards the plains, is gradually compressed by the upper strata, and the quantity of caloric, which had disappeared because of the expansion, is reproduced; the cold wind of the summit is heated again to blow in the valleys. This is a remarkable phenomenon in the mountains which separate Valais from Piedmont and Lombardy. From being very warm at the entrance to the Italian gorges of the Alps, the atmospheric current from the south is cooled by from 35 to 55 degrees in passing over Mont Rosa; it lets fall rain and snow in abundance; then, after having descended again on the opposite slopes, it brings to the

* *A Year's Travel in Central Arabia*, 2 vols.

peasants of Switzerland something of the burning climate of the tropics.*

As to the fearful tempests or *tourmentes* which occasionally surprise the traveller on high mountains or in the snowy plains, they may result from winds blowing from almost any point of the horizon. It is a terrible thing to be assailed by one of these phenomena. The white masses carried by the gusts of wind hide all surrounding objects. The unhappy people lost in this storm see neither the neighbouring slopes nor the sky above their heads, nor even the path beneath their feet. Deafened by the noise of the tempest, blinded by the powdery clouds which lash their faces, frozen by the snow which hangs in stalactites to their hair and changes their clothes into stiff and heavy masses, the travellers soon lose their way, and sink stupefied by the cold. Hundreds of corpses of men and horses, which have fallen here and there in certain passes of the Karakorum and the Himalayas, recall these terrible snow-storms, which have prevailed over these mountains. Accidents of the same kind are very numerous also on the *paramos* of India, Chili, Bolivia, and Peru. Even the Pyrenees and the Alps, where the most frequented passes are provided with hospices, where travellers surprised by the whirlwind of snows may take refuge, many unfortunate persons perish every year in these *tourmentes*.

The countries to the south of France have also to submit to the effects of a wind, which is a real scourge; it is the wind from the north-west, to which popular imagination has given the name of "master" (*mistral, magistraou, maestrale*). It is caused, like the alternate winds from the mountains, by the juxtaposition of two surfaces unequally heated. This aerial current is unhappily well-named, for its speed, comparable sometimes to that of the hurricane, suffices to uproot trees and throw down walls. "The *melamboreas*," says Strabo, "is an impetuous and terrible wind, which displaces rocks, precipitates men from their chariots, and strips them of their vestments and arms." The Gauls of the valley of the Rhône saw in it their most dreaded god; they raised altars and offered sacrifices to it; the Provençals considered it with "Durance," and the "Parliament" as one of their three great calamities. This wind makes itself especially felt in winter and spring, when the Cevennes, covered with snow, have become relatively very cold, and the sea-shores continue to be heated daily by the rays of the sun: then the masses of air roll in volumes from the

* Helmholz, *la Glace et les Glaciers*.

summit of the mountains, to replace the ascending current of the expanded atmosphere, which is formed above the region of the coast-line. During the night, it is true, the low lands situated at the base of the mountains lose their heat by radiation, and the afflux of cold air diminishes, to recommence on the morrow, when the sun warms the atmosphere of the plains anew. In summer, the difference of temperature is less between the shores and the desert escarpments of the Cevennes. The mistral is very feeble during this season, or it even entirely ceases. In various parts of the coasts of Spain, Italy, Greece, and Asia Minor, winds of the same kind, known under other names, descend in the same manner from the summit of the bordering mountains.

CHAPTER VIII.

ZONE OF VARIABLE WINDS.—STRUGGLE OF OPPOSING WINDS.—MEAN DIRECTION
OF THE ATMOSPHERIC CURRENTS.—LAW OF GYRATION.

BEYOND the changing limits where the trade-winds of the two hemispheres blow, commence the zones of variable winds. There the masses of air flow now in one direction, now in another, and apparently in a very irregular fashion. Sometimes a single wind directs itself incessantly during whole weeks towards one point of the horizon; sometimes the atmospheric currents which succeed each other make the tour of the compass in a few hours; at other times, again, the air remains calm between two meteorological regions where the winds move in opposite directions. Indeed the word *weathercock* has become a synonym of all that is unstable and versatile.

That which contributes to this disorder of the air in Europe, and in the other lands which are outside the zone of the trade-winds, is the inequality of the ground. The general currents which pass above a chain of mountains do not blow with the same regularity as in the plain. In fact, the winds must be all the more unequal in their successive gusts, the less even the surface is over which they blow. The same wind which moves over the seas with the uniformity of an immense river, departs from its regular pace as soon as it is interrupted in its course by inequalities of the soil. At the foot of the grand mountains of Switzerland, and especially in the environs of Geneva, where the surface relief is already very varied, the alternations which are produced in the force of the wind are such that the anemometer sometimes indicates a variation of intensity from single to triple. In the high gorges of the Alps it often happens, even during violent tempests, that the atmosphere presents at intervals the most perfect calm. To all the furies of the tempest there succeed for an instant silence and repose, then the hurricane recommences to blow with great violence. This is because the atmospheric currents, similar in this respect to the rivers of the ocean, do not direct themselves invariably towards the same point of the horizon, and move by successive oscillations now to the right, now to the left of the axis of their movement. In consequence, when we find ourselves placed on a point in the mountains which commands a view of the highest peaks, we must, according to the various directions which the aerial current takes, be by turns exposed to the fury of the tempest, or protected by some high summit on which the force of the wind is broken.* Even in countries but slightly varied in surface, or over plains covered with houses and woods, the wind does not blow in equal manner like the trade-wind of the seas. It advances by a series of gusts and blasts, each one of which represents a victory of the atmospheric current over an obstacle on the plain. Close to the ground the wind is always intermittent, while in the heights of the air it proceeds almost always with an equal and majestic movement like the current of a river.

The abrupt gusts of the lower strata of this ocean are thus only secondary phenomena, and in all the sudden turns of the winds which one might easily believe to have occurred by chance, the disorder is more apparent than real. Though the wind makes itself felt by turns from every part of the horizon, there, nevertheless, exist only two atmospheric currents in each of the temperate zones: that which comes from the pole to replace the expanded air of tropical regions, and that which flows back from the equator after being raised in the heights of space above the stratum of the trade-winds. In the northern hemisphere these two winds set out, one from the north, the other from the south; but in consequence of the rotatory movement of the earth, their direction is gradually changed, like that of the trade-winds. The wind from the north changes into a wind from the north-east, while the wind from the south ends by blowing from the south-east. Thus, as Dove remarks, the greater part of the aerial currents deceive the observer, because they do not come from the regions whence they appear to blow. The wind from the north-east is in reality much more the wind from the north than the mass of air whose direction is truly southern; in the same way, the wind from the south-east is truly the south wind, and that which seems to come from the south has the south-east as a starting-point.

Two great aerial currents thus dispute the extent of each terrestrial hemisphere from the pole to one of the tropics. Generally, all this space is divided into vast oblique bands, composed of masses of air flowing in opposite ways, some from the pole, and others from the equatorial regions. The bands move over the circumference of the

* H. de Saussure, *Voyages dans Les Alpes*.

globe, and in the same space, it is now the polar wind, and now the tropical wind which prevails. But a compensation never fails to be effected between these atmospheric currents, and the wind neutralized or repulsed in one part of the hemisphere soon makes itself felt at another point. While the strife exists between two masses of air animated by contrary movements, the vicissitudes of the conflict and the gradual preponderance of one of the winds result in temporarily modifying the direction of the air, and making the weathercock turn successively to the various points of the horizon. It is from the meeting of two regular winds that the apparent irregularity of all the atmospheric system results.

Though the strife between the two aerial streams, now at one point, now at another, does not cease, they are not, however, equal in force, and one of them always finishes by gaining the victory after a longer or shorter period of resistance. This wind of superior force is the returning current, descending from the heights of space to reach the level of the ground beyond the zone of the trade-winds. In fact, it is evident that in its circuit round the planet any one stratum of air must be much more expanded when it repairs from the torrid zone to the frozen regions, than it is on its return from the pole, after having been condensed by the cold; it occupies thus, in consequence of its temperature, a much greater space in the first journey. This is not all; the vapours with which the air of the equatorial zone is loaded, contribute to expand it still more, while the polar winds are relatively dry, and, consequently, much more dense. Thus the winds which come from the tropical zone, that is to say, the south-west winds in the northern hemisphere, and those from the north-west in the southern hemisphere, must have the preponderance, and blow during a more considerable space of time. It is thus at least in the temperate zone of the north, where the winds which are directed

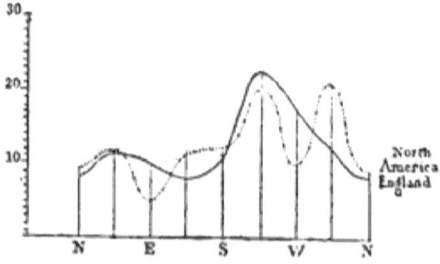

Fig. 108.—General direction of winds in England and North America. The total duration of the atmospheric currents for the year is represented by 100.

CONTEST OF OPPOSING WINDS.

towards the northern pole gain the victory, on an average, over the opposing winds.

As the atmospheric currents coming from the equator bend naturally towards the east, it follows that in the northern hemisphere most of the winds blow from the west. This is what we observe in North America, as well as in England. On the Atlantic coasts of France, the proportion between the winds which balance themselves around the western wind, and those which blow from directly contrary points of the horizon, is about three to two. The proportion would be much more favourable to the former if the chain of the Pyrenees, erected like a high barrier at the south of France, did not modify the direction of the atmospheric currents, and force them to make a detour by the Bay of Biscay, to bend again towards the east. At Cherbourg, in the open Channel, the difference between the winds

Fig. 100.—Map showing the general direction of winds in France.

from the west and those from the east is much greater. According to M. Liais, it is as seven to three. In the valley of the Saône and Rhône, the general movement of the winds is from north to south, as if the air were obliged to plunge in the kind of funnel formed by the Vosges, the Jura, and the Alps to the east; the heights of the Côte d'Or, Beaujolais, and the Cevennes, to the west. It is the same with every secondary valley. Thus the people of Valais scarcely know any winds but those from the east and west; in the high valley of the Rhône, the only winds which make themselves felt are those from the north and south.*

According to Kämtz, the mean direction of the wind in the whole of France is S. 88° W., that is to say, that the resultant of all the currents would blow from a point in the horizon situated at two degrees to the south of west. This direction of the wind explains perfectly why the large towns in France and the neighbouring countries tend in general to increase on the side of the west; they seek to breathe pure air. It is for this reason that the rich inhabitants of the great cities emigrate, from generation to generation, towards those portions of the suburbs which look towards the setting sun.

It is a remarkable fact, that the winds from the south-west increase in intensity in proportion as they approach the pole, while the winds from the north-east diminish gradually in force as they approach nearer the equator. The phenomenon is easily understood. The space traversed by the masses of air, coming from the south, is gradually restricted in the direction of the pole, and consequently the flow of the whole aerial river cannot be effected save by an acceleration of speed. The polar winds, on their side, traverse latitudes where the space opens wider and wider before them, and their force slackens gradually to the tropical zone, where they become the peaceful and regular currents of the trade-winds.

Already, for some centuries, savants have ascertained that, in the northern hemisphere, the succession of the winds is accomplished in a normal manner in the direction from south-west to north-east by the west and north, and from the north-east to the south-west by the east and south; this is a rotatory movement similar to that which the sun seems to describe in the heavens, when, after having risen in the east, it proceeds towards the west, developing its vast curve around the zenith. Aristotle made this observation more than two thousand years ago in his *Meteorology*: "When a wind ceases to blow, and gives place to another wind of a neighbouring direction, the change

* Tschudi, *Die Alpenwelt*.

takes place according to the path of the sun." Since the time of the great Greek naturalist many authors, whom Dove has taken the trouble to enumerate, have re-affirmed this fact of the regular rotation of the winds, which was besides known to sailors from time immemorial.

"When the wind veers against the sun,
Trust it not, for back it will run,"

is a seaman's adage. Nevertheless it is only in the nineteenth century that this meteorological phenomenon has been put beyond all doubt. Dove was the first to combine the scattered testimonies which confirm the popular idea, and transform the ancient hypothesis into a scientific certainty. For the future, it has become an incontestable fact, thanks to the savant of Berlin, that in the northern hemisphere the winds succeed each other most frequently in a regular order, which is indicated by the following formula:

S.W., W., N.W., N., N.E., E., S.E., S., S.W.

In the southern hemisphere the normal rotation of the aerial currents is accomplished in the opposite direction, that is to say, from north-west to south-east by the west and south, and from the south-east to north-west by the east and north:

N.W., W., S.W., S., S.E., E., N.E., N., N.W.

Thus in each of the opposite hemispheres the procession of the winds coincides with the apparent path of the sun, which for Europeans describes its daily course to the south of the zenith, and for the Australians passes to the north of this same point. Such is the regular order to which the discoverer has given the name of "law of gyration," and which is often and very justly designated by the name of "Dove's law." Thus the general winds themselves follow, in their succession, the same order as the little diurnal breeze caused by the relative position of the earth and the sun;* and it is perhaps owing to the support of these light breezes that the normal condition of the rotation of the aerial currents is established in space.

It is shown, by a great number of observations made in different parts of Europe, that the complete revolutions of the winds in the normal direction are much more numerous than those that occur in a retrograde direction. At Liverpool, London, Brussels, and Kharkov, the direct revolutions constitute, on an average, two-thirds of the total revolutions; in this respect, there is an almost perfect agreement between the atmospheric system of western and that of eastern Europe. In studying the partial revolutions, one does not always arrive at an

* See above, p. 261.

analogous result, because the direction of an atmospheric current often oscillates to the right and left of one point in the horizon, before describing a complete rotation in one direction or the other. Nevertheless, in order to guard oneself against all errors, it is important to study assiduously all the oscillations of the weathercock, for if such a complete gyration of the wind is not effected in the space of one month, the other kind can be completed in the space of a day. At Gnadenfeld, in Silesia, Kolbing observed a normal rotation, the duration of which did not exceed 16 hours, which is the length of a winter night.*

* Dove, *Loi des Tempêtes*, p. 92; Poggendorff's *Annalen*, lxii. p. 273.

BOOK II.—HURRICANES AND WHIRLWINDS.

CHAPTER IX.

AERIAL EDDIES.—CYCLONES OF THE EQUATORIAL REGIONS.—THE "GREAT HURRICANE."

It is probable that the wind is never propagated in a straight line. If it were it would be because it did not meet, in its course, any salient points of the surface of the earth, nor strike against any other masses of air, either at rest or moving in opposite directions. The atmospheric currents having always to strive against obstacles of this nature, must necessarily rebound to right or left, and advance by a series of eddies similar to those which the waters of a river form at the meeting of two currents. It is thus that a sudden wind raises the dust from the high road, or drives before it the leaves of the forest. In the same way, during the winter days, when unequal breezes chase each other in the atmosphere, the flakes of snow, in descending, describe long spirals, and the smoke which rises unrolls itself in circles of an ever-increasing diameter. The particles of air, like the heavenly bodies themselves, revolve as they move.[*] If two gusts of air meet at the entrance of a valley, and are continued in long eddies, the circular movement is continued from place to place, like a wave on the surface of the water, and the entire aerial mass is disturbed in its equilibrium.

In all the regions of the atmosphere, where two currents strike one another directly, or come in contact laterally, aerial eddies are instantly produced on the line of meeting, which move with extreme rapidity, and their vast whirls soon re-establish the equilibrium between the two masses of air. When these eddies have only a local importance they are known under the name of whirlwinds; but when their effects are felt over a great extent of country, the more general and more scientific designation of cyclone, proposed by Piddington, is

[*] Carus, *Natur und Idee.*

employed. This term can be equally applied to the hurricanes (in Caribbean, *aracan, huirauruean*) of the West Indies, to the *tornados* of the coasts of Africa, to the typhoons (ti-foong) of the Chinese Seas, to the revolving tempests of the Indian Ocean, and to the great gales of Western Europe. Still, we principally designate by the name of cyclone those whirlwinds which are developed according to a regular curve, either in the sea of the Antilles, or in the Indian Ocean, or more rarely in the Pacific Ocean.

Meteorologists have ascertained that the revolving tempests of the equatorial regions occur especially at the time of the reversal of the regular winds. Poey tells us that out of 365 hurricanes which have blown in the West Indies from 1493 to 1855, 245 (more than two-thirds) have taken place in October, that is to say, during the months when the strongly-heated coasts of South America began to attract towards themselves the colder and denser air of the northern continent.* In the Indian Ocean it is principally towards the vernal equinox, at the time of the change of the monsoons, and after the great heat of the summer, that the cyclones are most numerous. In the list of hurricanes in the southern hemisphere drawn up by Piddington and completed by Bridet, not a single cyclone is mentioned for the months of July and August; more than three-fifths of these phenomena have taken place during the three first months of the year. It is at this epoch of the change of the seasons that the powerful aerial masses, charged with electricity, engage in strife for the supremacy, and by their encounter produce those great eddies which are developed in spirals across the seas and the continents. Still the whirlwind never occupies in height more than a small part of the atmosphere. According to Bridet the mean height of the hurricanes of the Indian Ocean is rather less than two miles; and according to Redfield it is very rare that a cyclone would prevail at the same time at the level of the sea and at more than a mile above it. Ordinarily the revolving stratum of air is much less thick; occasionally it is even so thin that the sailors in a ship, whirled round by a cyclone, see above their heads the blue sky or the stars. Above this storm the winds follow their regular path.

These sudden movements of the air are perhaps, after the great volcanic eruptions, the most terrible meteorological phenomena of our planet, and we cannot be astonished that in the mythology of the Hindoos, Rudra, the chief of winds and storms, should have ended by becoming, under the name of Siva, the god of destruction and death.

* Poey, *Table Chronologique des Ouragans*, &c., 1862.

Some days before the terrible hurricane is unchained, nature, already gloomy and as if veiled, seems to anticipate a disaster. The little white clouds which float in the heights of air with the counter trade-winds, are hidden under a yellowish or dirty-white vapour; the heavenly bodies are surrounded by vaguely iridescent halos and heavy layers of clouds, which in the evening present the most magnificent shades of purple and gold stretching far over the horizon, and the air is as stifling as if it came from the mouth of some great furnace. The cyclone, which already whirls in the upper regions, gradually approaches the surface of the ground or water. Torn fragments of reddish or black clouds are carried furiously along by the storm which plunges and hurries through space; the column of mercury is wildly agitated in the barometer, and sinks rapidly; the birds assemble, as if to take counsel, then fly swiftly away, so as to escape the tempest that pursues them. Soon a dark mass shows itself in the threatening part of the sky; this mass increases, and spreads itself out, gradually covering the azure with a veil of a terrible darkness or a blood-coloured hue. This is the cyclone which falls and takes possession of its empire, twisting its immense spirals around the horizon. The roaring of the sea and skies succeeds to this awful silence.

The progress of the wind experiences much more resistance in the interior of continents than on the seas; but the phenomena which are produced there during hurricanes are not less terrible. Buildings which occur in the path of the storm are razed to their foundations, the waters of rivers are arrested and flow back towards their source, isolated trees are torn up and plough the earth with their roots, the forests bend as if they formed but a single mass, and give to the tempest their broken branches and torn leaves. Even the grass is uprooted, and swept from the ground. Innumerable fragments fly in the track of the hurricane like the waifs carried away by a fluvial or marine current.* Ordinarily, the action of electricity is added to the violence of the air in movement, to increase the ravages of the tempest. Sometimes the flashes of lightning are so numerous that they fall in sheets like cascades of fire; the clouds, and even the drops of rain, emit light; the electric tension is so strong that sparks have been seen, says Reid, to dart spontaneously from the body of a negro. An entire forest in St Vincent's Isle was destroyed without a single trunk having been overthrown. In the same way on the shores of Lake Constance in Europe, a great number of trees

* Audubon, *Birds of America*.

276   THE ATMOSPHERE AND METEOROLOGY.

which had remained upright in spite of the storm were completely stripped of their bark.

It is principally on the shores of islands and continents where the tempest has not yet been retarded by the obstacles of the ground that the effects of the storm are the most violent. It is there, too, that the greater number of human lives are destroyed in the general disaster; for then the ships always repair to the ports, and in many places of the coasts there are low lands which the waters suddenly rising inundate to a vast extent. Nevertheless, when the cyclone strikes against the mountains of a coast it cannot surmount them, and the regions situated beyond remain completely sheltered. Thus in the island of Réunion the hurricane only strikes one side of the island at the same time; too low to cross the mountains, it at first only devastates the plains situated on one side; but in its march

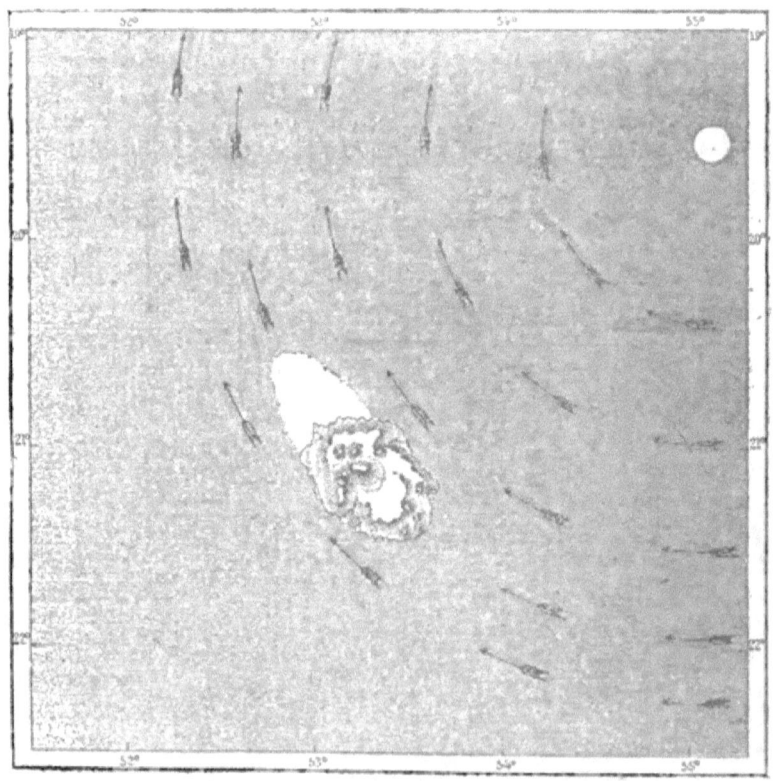

Fig. 110.—Calm during the hurricane at Réunion, Feb. 15, 1861.

across the sea the wind doubles the promontory that arrested it, and the ravages are instantly recommenced. Since the time of Columbus, the first European who contemplated the hurricanes of the Antilles, thousands of ships have been swallowed up during the revolving tempests of the tropical seas, either in the depths of the ports and roads, or in the seas that bathe the coasts of America, China, Hindostan, and the islands of the Indian Ocean. Such a cyclone as that of Calcutta in 1864, or of Havanna in 1846, has shattered more than 150 large ships in a few hours; such another catastrophe of the same kind, especially that which passed over the delta of the Ganges in October, 1737, drowned more than 20,000 persons in the rising waters.

In the midst of the ocean the dangers which ships run are less than in badly enclosed roads of the coast; but the sensations experienced by the seamen must be all the more lively, by their being completely isolated and lost in the awful whirlwind. Around them the daylight is darkened, and darker than night one might say, since the

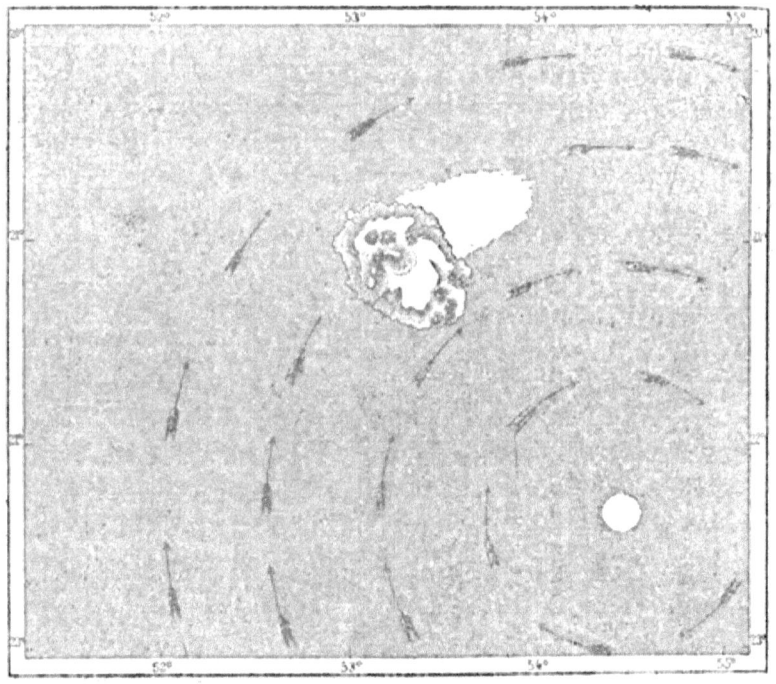

Fig. 111.—Calm during the hurricane at Réunion, Feb. 17, 1761

little light that still remains serves only to show the gloom. The winds which howl and whistle, the waves which dash against each other, the masts bending and breaking, the groaning of the timbers of the ship, all these numberless sounds are mixed and confused in a terrible despairing wail, drowning even the peals of thunder. The sea no longer rolls in large and mighty waves, but boils over like an enormous cauldron, heated by the fire of submarine volcanos. The low clouds creeping above the waters often emit a lurid light that one would say was the reflection of some invisible Gehenna; at the zenith appears, surrounded by darkness, a whitish space which sailors have named "the eye of the tempest," as if they really saw a fierce god in the hurricane who descends from the sky to seize and destroy them. When, in the middle of this terrible storm, the sailors accept the strife with the elements, and, defying death, seek to manœuvre and steer their dismantled ship without sails or masts, they certainly furnish a sublime example of human greatness.

Among the effects that certain hurricanes have produced, there are several which would seem quite incredible, if the genius of man could not by means of powder and other fulminating matters impress on the air a still greater rapidity and give it thus, though in very limited spaces, a force of destruction superior to that of the tempest. On the 26th of July, 1825, during the hurricane of Guadeloupe a gust of wind seized a plank an inch thick and sent it through the trunk of a palm tree 16 inches thick. In the same way in a lesser whirlwind which passed near Calcutta, a bamboo was hurled through a wall of a yard and a half in thickness; that is to say, the breath of air in movement over this point had a force equal to that of a six-pounder.* At St. Thomas, in 1837, the fortress which defends the entrance of the port was demolished as if it had been bombarded. Blocks of rock were torn from a depth of 30 or 40 feet beneath the sea and flung on shore. Elsewhere solid houses, torn from their foundations, have glided over the ground as if flying before the tempest. On the banks of the Ganges, on the coasts of the Antilles, and at Charleston, vessels have been seen stranded far from the shore in open plains or in forests. In 1681 a vessel from Antigua was carried up the rocks three yards above the highest tides, and remained like a bridge between two points of rock. In 1825, at the time of the great hurricane of Guadeloupe, the vessels which were in the road of Basse Terre disappeared, and one of the captains happily escaping, recounted how his brig had been seized by the hurricane and lifted out of the water,

* *India Review.* Dove, *Loi des Tempêtes.*

so that he had, so to speak, "been shipwrecked in the air." Broken furniture, and a quantity of ruins from the houses of Guadeloupe, were transported to Montserrat over an arm of the sea 50 miles wide. From the mountains of St. Thomas the immense black whirlwind was seen from afar to pass across the sea and over the islands of Porto-Rico and Santa-Cruz.

The most terrible cyclone of modern times is probably that of the 10th of October, 1780, which has been specially named "the great hurricane." Starting from Barbadoes, where neither trees nor dwellings were left standing, it caused an English fleet anchored off St. Lucia to disappear, and completely ravaged this island, where 6000 persons were crushed under the ruins. After this, the whirlwind, tending towards Martinique, enveloped a convoy of French transports, and sunk more than 40 ships carrying 4000 soldiers; on land the towns of St. Pierre and other places were completely razed by the wind, and 9000 persons perished there. More to the north, Dominique, St. Eustatius, St. Vincent, and Porto-Rico were likewise devastated, and most of the vessels which were on the path of the cyclone foundered with all their crews. Beyond Porto-Rico the tempest bent to the north-east towards the Bermudas, and though its violence had gradually diminished, it sunk several English warships returning to Europe. At Barbadoes, where the cyclone had commenced its terrible spiral, the wind was unchained with such fury, that the inhabitants hidden in the cellars did not hear their houses falling above their heads; they did not even feel the shocks of earthquake which, according to Rodney, accompanied the storm. The rage of man was arrested before that of nature. The French and English were then at war, and all the ships which the sea swallowed up were laden with soldiers seeking to destroy one another. At the sight of such ruin the hatred of the survivors was calmed. The governor of Martinique caused the English sailors, who had become his prisoners in consequence of the great shipwreck, to be set at liberty, declaring that in the common danger all men should feel as brothers.

CHAPTER X.

SPEED OF THE REVOLVING MASSES OF AIR.—SPEED OF THE CYCLONE.—FALL OF THE BAROMETRIC COLUMN.—IRREGULARITIES OF THE WIND IN THE PATH OF THE CYCLONE.

It is not yet known what degree of swiftness the masses of air carried by the cyclones can attain, for it is in the upper regions of the atmosphere, where the medium only offers a feeble resistance to the aerial currents, that the storm-wind must have its greatest rapidity. And it does not suffice to ascertain the progress of the particles of air immediately at the level of the ground, or even slightly above it, to form an idea of the speed at which the atmospheric mass carried by the hurricane moves. In one of his ascents Mr. Coxwell made a journey of 68 miles in 60 minutes, while below him the instruments indicated a speed of hardly 14 miles in the same interval. Another time Mr. Glaisher moved at 15 miles per hour, while at the Greenwich observatory the same sheet of air only advanced 500 yards. How great, then, is the speed of the cyclone at a certain height above the ground when on the earth strewn with obstacles it progresses at the rate of 50 yards per second, or 100 miles per hour, four times the speed of our locomotives! This fearful rapidity of the air at the surface of the ocean, and the friction of the aerial particles which results, explains perfectly, as Cicero remarked 2000 years ago, why the temperature of the water rises during storms.*

As to the pressure exercised by the aerial current which moves with such speed, it is truly formidable. In a memoir on the Construction of Lighthouses, Fresnel estimated the strongest pressure of the wind at 616 lbs. per square yard, but it is very probable that in a number of hurricanes this figure has been greatly surpassed. Not to mention the effects produced by the great cyclones of the tropics, a number of cases have presented themselves in the temperate zone where the pressure exercised by the wind on a space of little extent was much greater than meteorologists had foreseen. Thus, to cite but one example, the storm of the 27th of February, 1860, coming from the

* *De Natura Deorum—Zeitschrift für Erdkunde,* March, 1864.

west, and plunging in the plain of Narbonne by the strait where the canal and railroads of the south pass, was violent enough to force off the rails and partially overturn two trains which it struck crossways, between the stations of Salces and Rivesaltes. According to the engineer, Mathieu, who probably gives, it is true, too high an estimate, the pressure necessary to overturn certain carriages must have been 952 lbs. per square yard of surface.*

The masses of air which revolve not far from the central part of the cyclone are the only ones which attain the considerable speed of 60 and 90 miles per hour. As to the movement of the whole of the storm on the surface of the earth, it is naturally very slow in comparison to the circulatory movement of the aerial particles around their axis. The greatest speed of translation which has been observed is that of the hurricane in the month of August, 1853, which, after having advanced at the rate of 20 miles an hour from the Antilles to the bank of Newfoundland, increased gradually in speed, and ended by exceeding 56 miles an hour. Most of the cyclones of the Antilles move on an average from 12 to 18 miles in the same space of time; but there are some too, especially among the typhoons of China, which advance so slowly that several writers have considered them as revolving on the same spot. At the end of the month of February, 1845, a hurricane which originated near the Mauritius traversed the Indian Ocean with an average speed above 2 miles per hour, while a ship, the *Charles Heddles*, placed at about 56 miles from the axis of the storm, described immense spirals around this

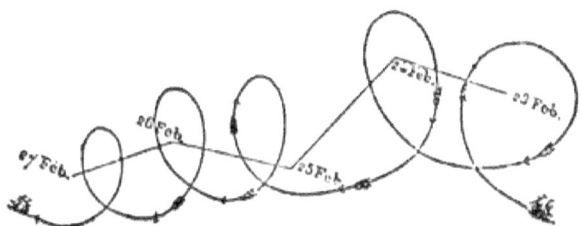

Fig. 112.—Spirals made by the vessel "Charles Heddles."

changing point. In five days it made five complete revolutions in the midst of the sea, and though in this fantastic voyage it must at least have traversed 1500 miles, nevertheless, when it was finally delivered from the grasp of the cyclone, it was only at 410 miles from the point of departure. The vessel had revolved like a top on the

* Eugène Flachat, *Traversée des Alpes*.

surface of the ocean. According to Bridet,* the speed of translation in the hurricanes of the Indian Ocean is comprised between the extremes of one mile and 20 miles an hour.

The movement of the cyclone has the effect of hollowing into a funnel all the central part of the whirlwind, and hurling the masses of air towards the circumference of this enormous wheel which turns in the atmosphere. It is thus that in the rivers, and even the smallest tributaries, the eddies are always depressed in the centre, because of the centrifugal force which carries the waters along in a circle. The diminution of the aerial column makes itself instantly felt, by a corresponding diminution of weight, and the mercury sinks in consequence, as soon as the hurricane commences to be formed in the high regions of the atmosphere. The storm which is approaching thus announces its proximity, and those whom it threatens can take their precautions so as to escape entirely from the disaster, or so as to diminish its effects. The sailors whose vessel is anchored in a sure port, double their moorings; those who are lying in an open road, exposed to the fury of the winds, as at Réunion, hasten to obey the signal gun, and fly to the open sea so as to withdraw from the centre of the hurricane. The barometer has been seen to fall by $1\frac{1}{2}$, 2, and even $2\frac{3}{4}$ inches,† that is to say, nearly a tenth of the total height of the mercury, and each of these perturbations has not failed to be the signal of a storm all the more terrible the higher the barometer had previously risen. At times the rarefaction of the atmosphere is accomplished in such a sudden manner, that the air contained in the houses suddenly expands, explodes, so to say, and hurls windows and doors far away. For this reason, says Fitzroy, the habitations are left open in certain places to avoid such accidents.

In the sea, the waters rise to a greater or less height in consequence of the lessening of the atmospheric pressure, and move with the centre of the cyclone; thus a "tempest-wave" is raised, whose force is added to that of the formidable surf which the wind has excited. This is the principal cause of those terrible tidal "races," no less dangerous than earthquakes which roll over the neighbouring coasts. During the hurricane of Barbadoes, in 1831, the waves which broke against the northern promontory of the island were 72 feet higher than the mean level of the water. At the great cyclone of Calcutta, in October, 1864, the Hooghly rose 22 feet all along the lower part of its course, and inundated several islands. More recently

* *Etude sur les Ouragans de l'Hémisphère Austral.*
† On board the *Duke of York*, in 1833, at the mouth of the Hooghly.

ROTATION OF CYCLONES.

still, in the great hurricane which devastated St. Thomas, a wave driven by the wind rushed over the small island of Tortola, committing such ravages that, according to an absurd legend propagated by terror, the entire island was swallowed up. It is certain, too, that the water of the sea can be drawn in in greater or less quantity by the vacuum which is formed in the midst of the whirlwind; this has occurred many times, and especially in Barbadoes. Reid saw showers of salt-water fall at a great distance from the shore in the interior of the island, and destroy all the fresh-water fish in the lakes and streams.

The circular movement of the cyclones does not occur indifferently in one direction or the other. Like the regular phenomena of the winds, these terrible storms, as well as all the other great atmospheric perturbations, conform to laws, and their progress can therefore always

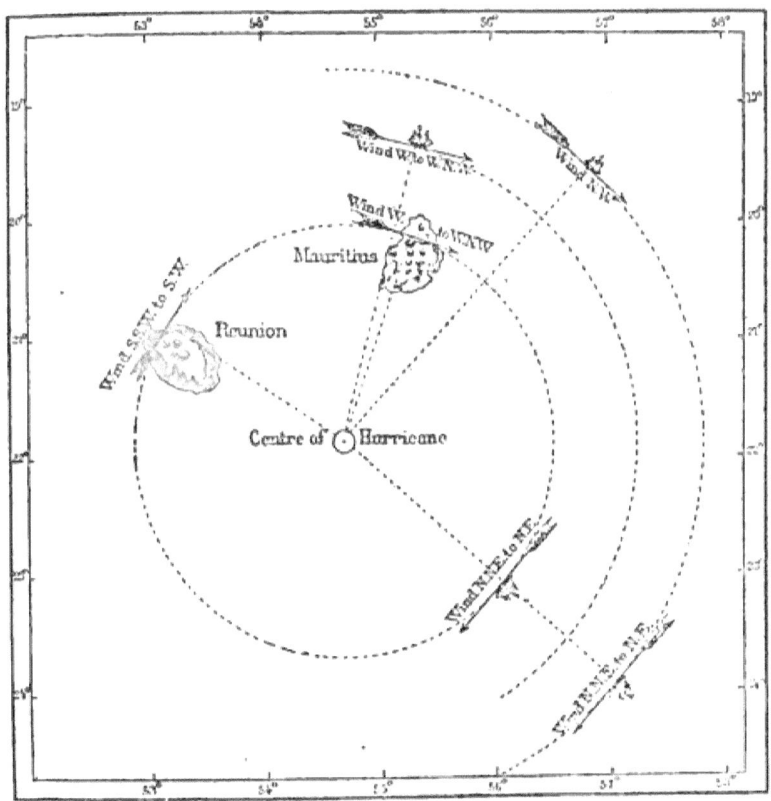

Fig. 113.—Cyclone in the Indian Ocean in Jan., 1852.

be foretold by sailors. In the northern hemisphere, the revolving storms of the tropics constantly blow from the south to the north by the east, and from the north to the south by the west; in the southern hemisphere, the path taken by the whirlwinds is in the opposite direction, and the spirals of the wind are uniformly developed by the south, the west, the north, and the east. Such is the law discovered and brought to light by the labours of Reid, Redfield, Piddington, Bridet, and other savants. Thus, winds from all parts of the horizon blow at the same time round the circumference of the cyclone; one ship is pursued by a furious wind from the east, while at 50 miles distant another vessel is sunk by gales coming from the west. And during all these tumults of warring elements, it sometimes happens that at

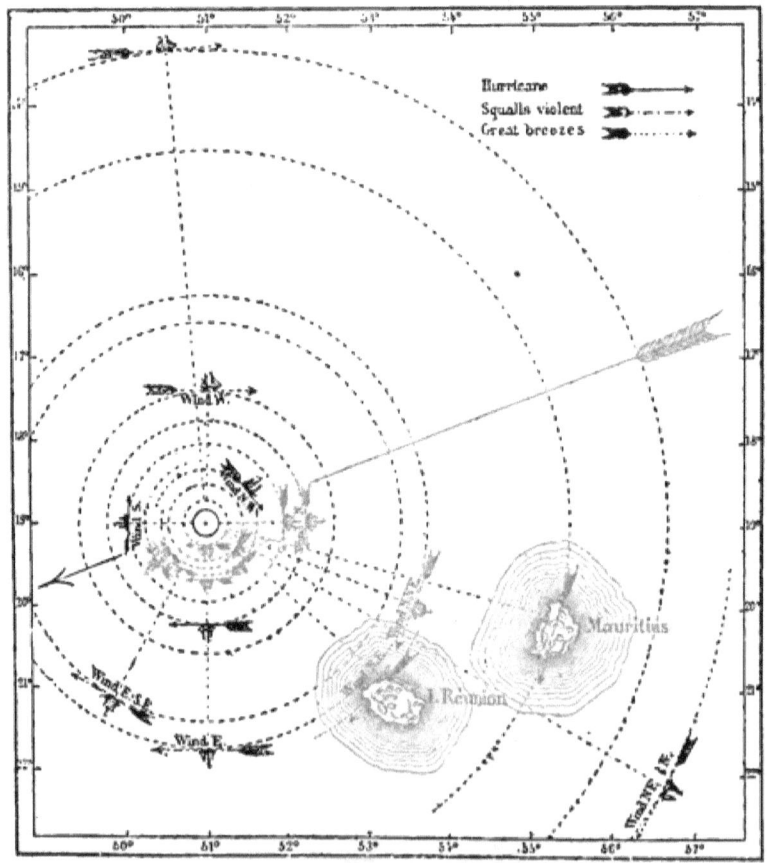

Fig. 114.—Cyclone in the Indian Ocean in Feb., 1860.

the very centre of the hurricane the atmosphere remains perfectly calm; a terrible peace, a formidable silence, reign in the changing circle formed by the raging whirlwind of the tempest.

If the cyclone turned round in its place, the wind would blow exactly in the direction of the tangent over the whole course of the storm; but it is not thus because of the double movement of the hurricane; while revolving it moves on, and consequently the direction of the wind must be the result of the two forces which bear it along. Let the entire whirlwind be directed towards the west, and

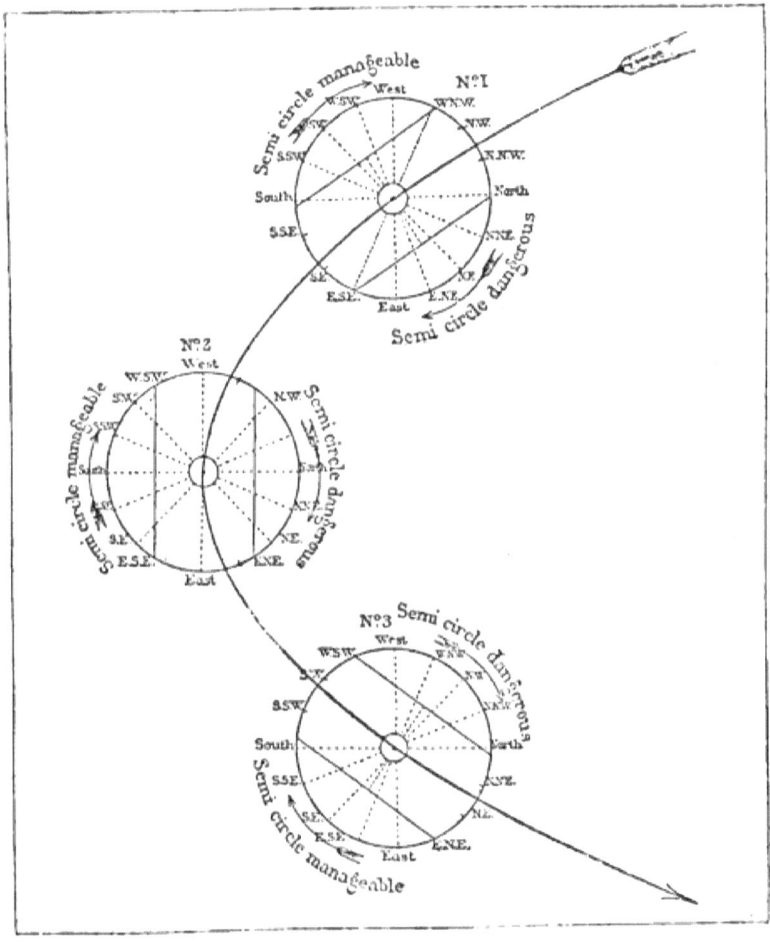

Fig. 115.—Parabola described by a hurricane (after Bridet).

the normal speed of the wind of the tempest, which blows in the same direction over the periphery of the cyclone, will be augmented by the speed of the storm itself. In return, the wind which will blow towards the east, will be partially neutralized, and along all the outline of the circle, the direction and the speed will be modified according to proportions rigorously established by calculation. These are the modifications to which the successive winds along the outline of the tempest are subject, and which often render the cyclones difficult to recognize in the regions of the temperate zone, where the speed of rotation of the storms is considerably diminished. Under the tropics, where the whirlwind, being still restricted, is in its primitive force, we remark the less this inequality of these partial winds of the hurricane. It is, however, important enough to be recognized by mariners. One half of the disk of the tempest is called by them "dangerous semi-circle," and the other "manageable semi-circle." Now, this part of the hurricane, which the great violence of the winds renders dangerous, is always found on the side of the cyclone where the wind proceeds in the same direction as the storm. That half of the disk where the wind adds its own speed to that of the movement of translation is in the northern hemisphere, to the right of the trajectory of the revolving circle; in the southern hemisphere, it is to the left.* The accompanying figure gives an idea of the contrast which occurs between the two sides of the hurricane on the path that it traverses in the Indian Ocean.

* Marié Davy, *Mouvements de l'Atmosphère et des Mers*.

CHAPTER XI.

SPIRAL OF THE HURRICANES IN THE TWO HEMISPHERES.—THEORY OF CYCLONES.
—NAUTICAL INSTRUCTIONS TO AVOID HURRICANES.

At their departure from the tropical regions, where they fell into collision with the trade-winds, or the monsoons, the greater part of the cyclones of the new world proceed first towards the northwest, parallel to the line of the Antilles, or else along the shores of Columbia, and Central America; then, turning back, like a billiard-ball that rebounds in the opposite direction from the impulse received, they follow the coast-line of the United States, describing in the air an orbit corresponding with the course of the Gulf-stream.

In the southern hemisphere the phenomenon is inverted; the cyclones of the Indian Ocean take their origin to the south of Hindostan, and move to the south-west towards Réunion, Mauritius, and Madagascar, then turn abruptly in a south-westerly direction towards the Antarctic seas. The spiral movement of the wind in this great tourbillon is effected from west to east by the north, that is to say, in the same direction as the hand of a watch. The movement is opposite to that which is taken by the hurricanes of the northern hemisphere.

What is the cause of the cyclone itself, and whence comes this sudden change, which occurs in its direction towards the exterior limit of the trade-winds? According to Dove, this is the explanation of these phenomena :—

When enormous quantities of warm air ascend over the deserts of Asia and Africa, these expanded aerial masses must spread laterally. Those which are carried over the North Atlantic in a westerly direction, contrary to that of the earth's movement, meet the returning current, which flows from the south-west to the north-east, in the opposite direction of the trade-winds. From this results a conflict between the two atmospheric currents; a whirlwind of air is propagated in spirals in the north-westerly direction, which is the result of the two forces at issue. At the same time the revolving mass

288   THE ATMOSPHERE AND METEOROLOGY.

descends obliquely towards the surface of the sea, and being compressed to the right by the trade-winds, it continues to advance towards the north-west. Arriving outside the tropics, the hurricane is no longer under the lateral pressure of the north-east wind; it has a free path before it, and under the influence of the earth's rotation, it bends with a graceful curve in a northerly direction, then in that of the north-east. At the same time, the storm which has just entered the temperate zone gradually enlarges the diameter of its spirals, and consequently loses its violence in proportion as it advances towards the pole. Thus the hurricane of 1839, whose breadth was about 300 miles, when it crossed the Antilles, extended to 500 miles

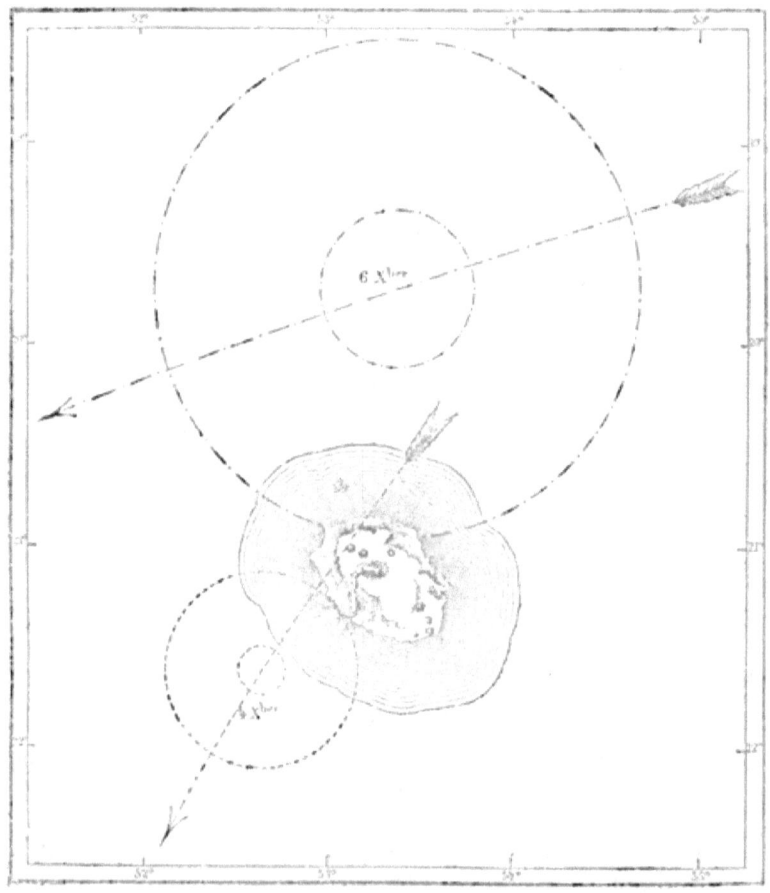

Fig. 116.—Simultaneous Cyclones experienced at Réunion, December, 1824.

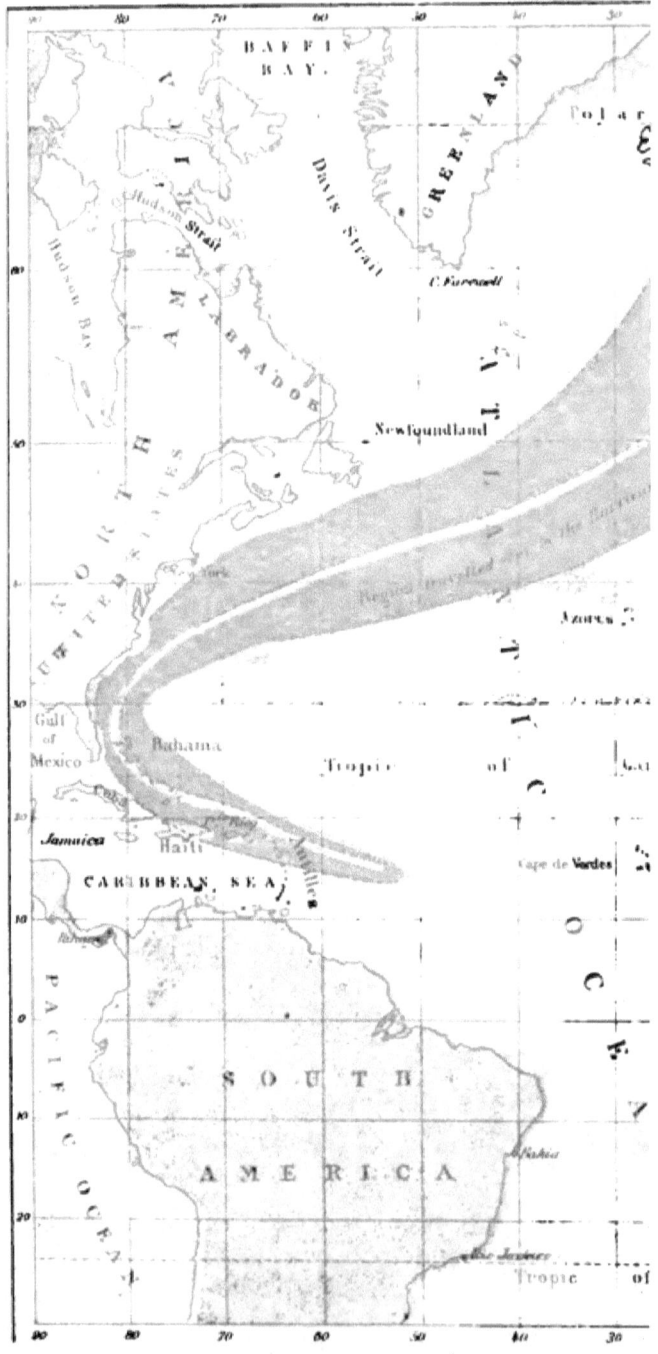

above the sea of the Bermudas, and about the 50th degree of north latitude, it did not occupy a space less than 1750 miles; but at the same time its destructive effects diminished in proportion to its expansion.

The same wind, which has just razed a town in the Antilles, and broken ships like playthings, sometimes contents itself, when it arrives at the Irish coasts, with uprooting a few trees and overturning some already trembling rocks.

Such is the theory proposed by Dove, and which seems the most probable, at least, for the hurricanes of the Atlantic. As to the cyclones of the Indian Ocean, they are, perhaps, produced by the conflict of the south-easterly trade-winds, and the monsoon, which tends towards the continent of Africa. M. Bridet sees in them only the result of the meeting of two winds, one from the equator, the other from the southern hemisphere. That from the equator, participating in the great angular speed of this part of the globe, deviates towards the east, in proportion as it advances towards the tropic of Capricorn; the south wind, carried less rapidly around the earth, deviates, on the contrary, towards the west, and from these two deviations in opposite directions there results, at the meeting of the winds, a revolving movement in the direction from east to west by the south. On an average, the cyclones of the Indian Ocean have a diameter of from 250 to 300 miles in the commencement of their course, from 430 to 560 miles towards the middle, and from 560 to 700 towards the end; their influence is sometimes felt as far as 1200 miles from the axis of the storm. It is true that two or more cyclones often follow one another at a little distance; lateral eddies accompany the principal whirlwind in the same way as occurs on the surface of the sea, beside the great revolving funnel, formed by the meeting of contrary waters, many circles of the second order are hollowed out. Bridet has collected numerous examples of these simultaneous cyclones.*

Local obstacles, such as plateaux and mountain-chains, may likewise cause hurricanes, when the aerial masses dash directly against them. Thus, in the Bay of Bengal, at the time of the change from the north-east to the south-west monsoon, the latter strikes against the mountains of Arracan, and in consequence of this shock a sudden cyclone occurs, which turns back towards the north-west, and traverses the whole of Bengal and the northern provinces of Hindostan as far as the Hindoo-Koosh. It is possible that the typhoons of the

* *Etude sur les Ouragans de l'Hémisphère Austral.*

290   THE ATMOSPHERE AND METEOROLOGY.

Chinese sea owe their origin to similar causes; in this case they would be nothing else than deviating monsoons, transformed into hurricanes,

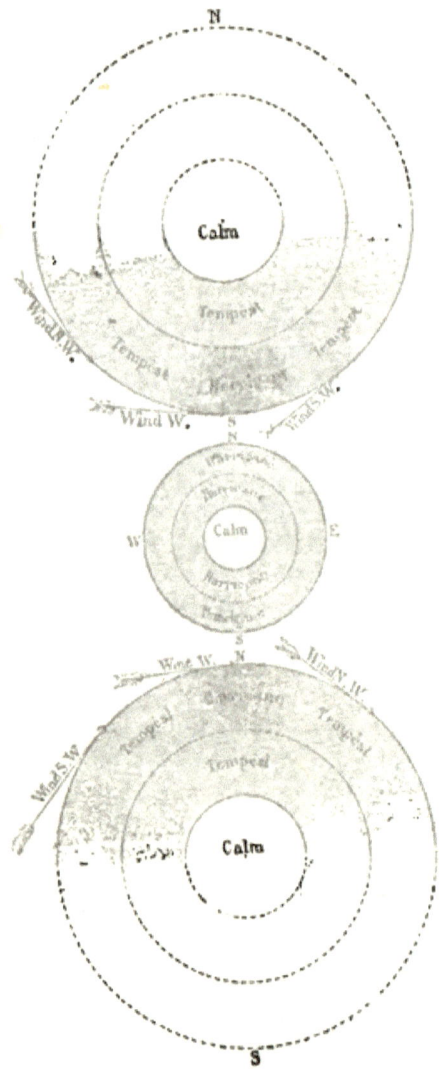

Fig. 117.—Direction of Cyclones on the surface of the earth.

because of the obstacles opposed to them by the mountains of the Philippines and Formosa. Besides all these hilly countries of different sizes and forms, which are scattered over that part of the Pacific

Ocean, and are separated from one another by unequal and tortuous straits, cannot fail greatly to disturb the normal condition of the winds,* and to produce a great number of storms and hurricanes often confounded under the general name of typhoons. On the other hand, on the Eastern Pacific, where the trade-winds blow with so much regularity, hurricanes properly so-called are very rare. They have only been observed on the eastern coasts of Mexico.

While the cyclone developes its vast curves in the equatorial regions, the entire whirlwind must lean forward, for the upper strata carried away in the hurricane find much less resistance in the air than the lower strata find above the ground and the surface of the sea. The whole of the storm may, therefore, be compared to an immense wheel, revolving horizontally over the globe, and pressing the earth most strongly with its anterior part. Nevertheless, in extending themselves in large spirals in the two temperate zones of the north and south, the hurricanes become gradually subject to such modifications in opposite directions, and present apparent irregularities so considerable that they seem at first to obey other laws. Instead of leaning forward, one would say, on the contrary, that a real vacuum, incessantly enlarging, was opened on this side of the whirlwind. Thus, as is proved by more than 300,000 observations made in the North Atlantic, on board American, English, and Dutch ships, and carefully compared by Messrs Andrau and Van Asperen,† the winds from the north, which have passed the thirtieth degree of north latitude, are almost always wanting in the spirals of cyclones. In proportion as the meteor is developed towards the pole, the tranquil zone of the hurricane increases. The winds from the south and from the east diminish by degrees in frequency and in intensity, they then disappear completely. Finally, from the 50th to the 60th degree of latitude the rotation of the cyclone is represented only by winds from the north-west, west, and south-west. One might say that only half the hurricane remains. To the south of the equator similar phenomena occur in inverse order, and every successive curve in the spiral of the storms presents in its southern convexity a greater or smaller break according to the height of latitude. Fig. 117 may explain the modifications which the cyclones experience while proceeding from the tropical regions towards the poles severally.‡

The fact that in the northern hemisphere the partial winds of the

* See above, p. 256. † *De Wet der Stromen.* 1862.
‡ *Mittheilungen von Petermann,* t. xi. 1862.

hurricane are always stronger on the right side of its path, and in the southern hemisphere always stronger on the left, is not sufficient to explain this astonishing contrast between the two halves of the disc of the cyclone. M. Andrau and other Dutch savants have attempted to explain this apparent anomaly. Taken altogether, the hurricane may be considered, they say, as a disc revolving rapidly around its axis. Its natural tendency is to move incessantly in the same plane of rotation, and only the intervention of considerable force can make it incline in one or the other direction. It is true that at the point where it originates over the equatorial seas the cyclone leans more or less strongly towards its source; but in proportion as it moves towards the pole revolving round an imaginary axis that remains always parallel to itself, it must necessarily lean more and more backward, in consequence of the curvature of the globe. While the southern part of the hurricane still sweeps over the waves or the plains, the other part rises gradually to a great height in the atmosphere. Soon the upper winds of the tempest no longer make themselves felt at the level of the soil, and are only indicated by the fall of the barometric column and by the clouds which we see hurrying after each other at a great height in the sky. Towards the 50th degree of latitude to the north or south of the equator, the cyclones elevated half-way only touch the earth by the winds of their lower extremity. These winds are the same in the two hemispheres, they blow equally from the north-west, west, and south-west; but on each side the gyration is accomplished in the opposite direction.

Piddington, Redfield, Bridet, Lartigue, and other learned meteorologists have drawn up rules of general conduct for mariners surprised by hurricanes, which, when they are followed in time, may save the threatened ship. Warned by the barometer of the approach of the cyclone, the captain must be very careful not to fly at full speed before the storm, in the vain hope of escaping the danger. By proceeding in this way, as terror would counsel him, he would rush precisely into the midst of the tempest and expose his ship to all the fury of the wind and the surf. To escape its violence he ought to manœuvre, so as to tend obliquely towards the circumference of the storm as far as possible from the central part, where the wind blows with all its force. Unhappily, whatever may be the science of the seaman and his knowledge of the seas which he navigates, it is often very difficult for him to know beforehand from which side the winds will approach, and what is exactly the orbit which the centre of the cyclone follows across the seas. Nevertheless, if he hesitates too long,

he may suddenly find himself within the fatal circle, and be lost with
his ship, from having lacked the necessary boldness. In the high
latitudes of ocean it is easier to make a decision, and escape from
the cyclone, since the sea is open in the direction of the pole, and the
sailor has not to dread being completely enclosed in the midst of a
circle of tempests. It is behind him that the lower part of the
immense wheel ploughs the waves, before him the ocean is open, or at
least the winds which traverse its surface are produced by local causes,
and do not belong to the terrible storm. Only at very rare intervals
is the upper part of the cyclone brought down to the surface of the
water by violent atmospheric counter-currents coming from the polar
regions. In thirteen years the Dutch savants have only observed
two cases of this nature.

Thus the hurricanes themselves, like the other manifestations of life
on our globe, have a regular course, and mathematicians can attempt
to calculate the orbit of these terrible phenomena over the face of the
earth. It is by conforming to laws and following spirals traced
beforehand that the revolving tempests are propelled from the
equinoctial zone to the temperate regions. Far from causing by their
violent spirals a permanent disturbance in the air, they, on the contrary, only re-establish the equilibrium between the unequal waves of
the atmospheric ocean. Still more, they aid conjointly with the monsoons and the counter trade-winds, to maintain the astronomical
equilibrium of the planet. Thus, as Dove remarks, the continual
friction of the trade-winds, which the terrestrial rotation causes to
deviate incessantly towards the west, would doubtless end by retarding
the movement of the earth around its axis, if other aerial currents
proceeding in an opposite direction did not counterbalance the
retarding causes, and accelerate on their part the rotation of the earth
from west to east. Slight as may be the breath of wind compared to
the force of projection which causes the planet to revolve, it does not
the less contribute to the movements of the globe and to its harmonious circles in the concert of the heavenly bodies.

CHAPTER XII.

EDDIES OF TEMPESTS.—WHIRLWINDS.

The atmospheric movements called tempests or gales by seamen differ from the cyclones by their slighter intensity, but are more numerous. In certain parts of the ocean, especially in the North Atlantic, they are so frequent, that during some months of the year we may expect a tempest once every two days. This is shown by the accompanying map, every rectangle of which indicates the number of

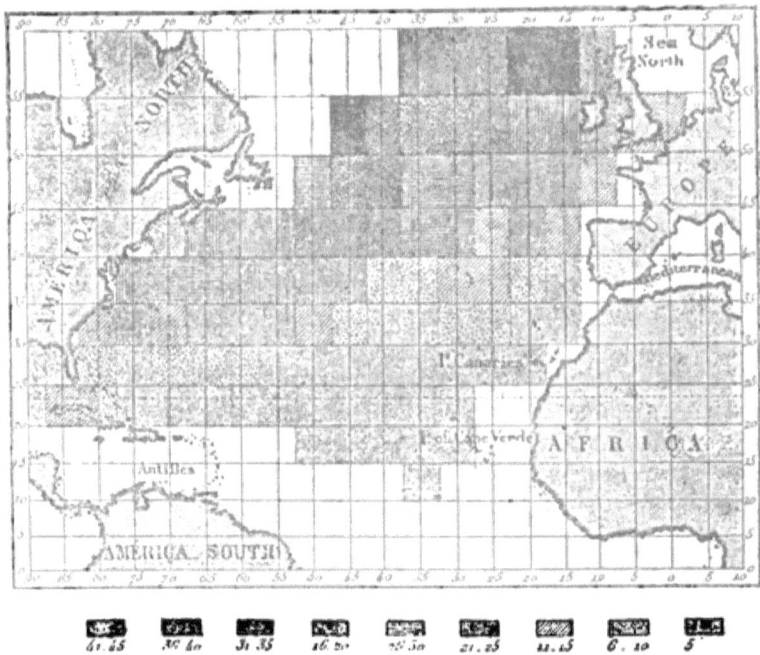

Fig. 118.—Tempests of the North Atlantic in December, January, and February.

tempests by a different tint. All these gales are propagated in spirals analogous to those of the hurricanes. Storms of winter or tempests of summer originate to the right or left of the Gulf-stream, and are developed in gyrations, caused by the movement of the earth itself.* There are likewise local cyclones, revolving only over a single country like France or England or even in a single valley; we might cite numerous examples of similar tempests which in a limited space have been scarcely less destructive than the hurricanes of the Antilles.† Often when we contemplate the sky above our heads, we

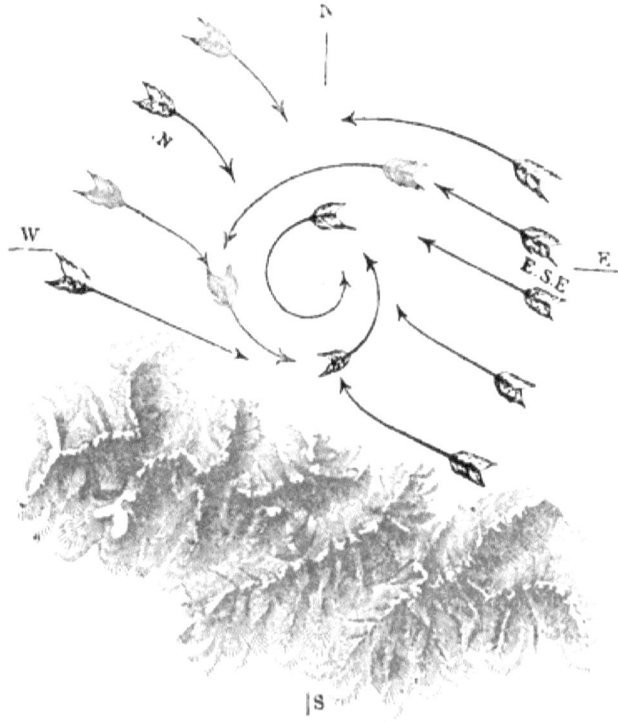

Fig. 119.—Storm in the Pyrenees; after Lartigue.

see clouds whirling under the influence of two hostile currents and approaching one another only to withdraw again. But it is principally by ascending the side of a mountain that one can witness the curious sight presented by the conflict of two masses of air, which dive into a valley and describe a more or less rapid eddy with their clouds or mists. From the top of the headlands of the Pyrenees the meteorologist Lartigue has observed a great number of these circular

* Sonrel, *Nouvelles Météorologiques*, March, 1868. † Fitzroy, *Weather-book*.

winds, similar to the circles which the water of a river describes above a rock.*

As to whirlwinds, properly so called, they are phenomena of small importance compared to the cyclones; but, like them, they are due to the encounter of two, more or less considerable, masses of air, which strike against each other obliquely. Still they do not turn invariably in one direction for each hemisphere, for they are not caused,

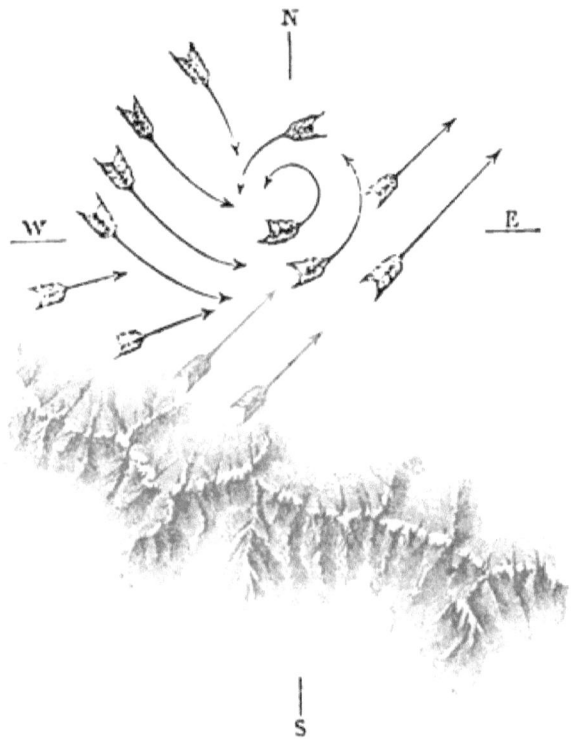

Fig. 120. - Storm in the Pyrenees; after Lartigue.

like the hurricanes, by the strife of two regular winds, but may arise from the conflict of all the currents of air, either normal or variable, which traverse the surface of the earth. Observers have seen in the same regions whirlwinds which revolve from the north to the south, some passing by the west, and others by the east. During a tempest there may even form on each side of the atmospheric current, as on the shores of a fluvial current, a series of eddies revolving in the contrary

* Lartigue, *Essai sur les Ouragans et les Tempêtes*.

direction, and sometimes with sufficient speed to deserve the name of whirlwinds. In the full whirl of the cyclone, the shock of the gusts

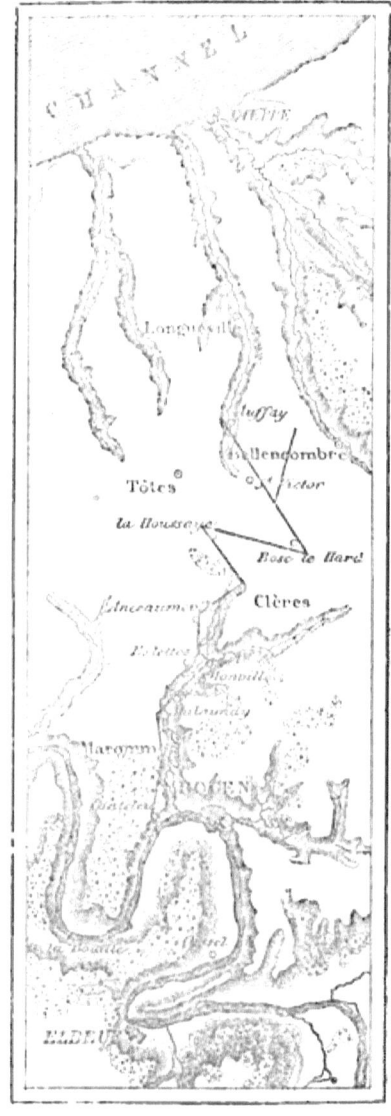

Fig. 121.—Hurricane of Monville.

of wind must likewise produce secondary eddies, moving with extreme speed, now in one direction, now in another. If it were not so, we should not be able to understand how in the very centre of the hurricane the effects produced by the wind differ in so remarkable a manner in a space of small extent. Thus, according to Reid, it has often been ascertained that during the cyclones of Mauritius lofty houses half in ruins already, were not even shaken by the storm, while solid buildings beside them were completely overthrown and destroyed.*

Isolated whirlwinds are sometimes propagated with a rapidity as great as that of the hurricanes, and may cause similar disasters. The whirlwind which passed over Malaunay and Monville on the 19th of August, 1845, was not more than 33 to 44 yards wide in certain places, and in its greatest breadth it hardly attained the third of a mile; notwithstanding which it committed the most terrible ravages, and the inhabitants of that part of Normandy long preserved the fearful memory of it. About one o'clock in the afternoon, after an oppressive day, during which the barometric column had suddenly fallen from 29·9 to 27·8 inches, some sailors saw the whirlwind forming over the Seine at the foot of the high cliffs of Canteleu. Like an inverted pyramid, blackish at the base and red at the summit, the whirlwind swept the waters with its point and then rushed into the valley of Maromme. It did not advance in a straight line, nor by elongated curves, but by abrupt deviations to right and left, like the zigzag of lightning. Through the woods which were on its path, it traced wide roads over trees overthrown, shattered, and reduced to ruin; then approaching successively three great silk manufactories of Monville, it twisted them in its spirals, and struck and destroyed them. After having heaped up all these ruins, under which perished hundreds of workmen, the whirlwind opened an avenue in the ruins on the plateau of Clères, then divided into two branches and ascended into space, carrying with it all kinds of objects, planks, slates, and papers, which fell down again near Dieppe at distances varying from 15 to 24 miles from the place of the catastrophe.† It is evident, according to all accounts, that electricity played a very great part in the whirlwind of Monville.

These phenomena, as we can understand, produce different effects according to the region that they traverse. Those which pass over forests break the trees or even twist them in various directions. Others which traverse large prairies, such as the pampas of Buenos-Ayres, the steppes of Turkestan, and the grassy countries of Central Africa, raise

* Lartigue, *Essai sur les Ouragans et les Tempêtes*, p. 89.
† Eugène Noël, *Documents Inédits* ; Dagnin, *Traité de Physique*.

myriads of locusts in their tourbillons, and carry them either to other parts of the continent, where these insects instantly devour all the crops, or towards the ocean, where they are swallowed up. Some-

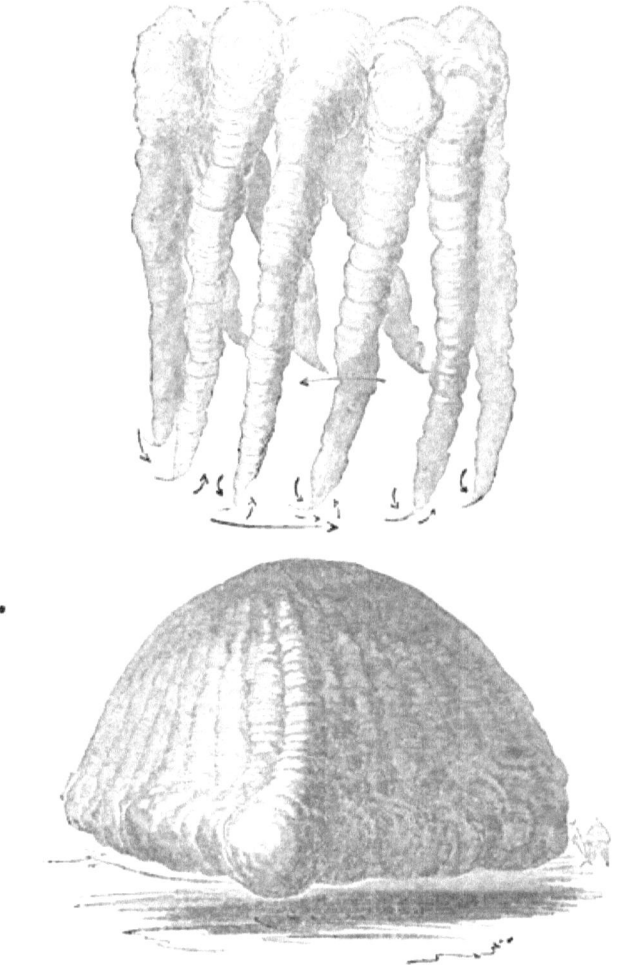

Figs. 122, 123.—Whirlwinds of Dust.

times the navigators encounter, at considerable distances from the coast of Africa, real clouds of them, that the tempests have raised from the ground and then consigned to the north-east trade-winds.*

In the deserts of the Sahara, Arabia, Khorassan, India, and South

* Lartigue, *Système des Vents*, pp. 70, 71.

America, the winds raise enormous quantities of dust, and cause them to revolve in space. At Buenos-Ayres the whirlwinds of 1805 and March, 1866, were powerful enough to render the atmosphere as black as night, and to stifle pedestrians in the streets; after the passage of the storm, the rain which fell showered mud upon the ground. Sometimes the masses of dust are columns revolving and dancing in immense circles like the genii of the air; sometimes, too, they are enormous cupolas whirling in space covering hundreds and even thousands of yards in breadth, and developing their ellipses for days together and to great distances. These whirlwinds render the atmosphere completely dark and irrespirable. In order not to be stifled, travellers are obliged to shut themselves up in all haste in their tents and to throw themselves down with their faces to the earth, so as to form a rampart of their own bodies against the storm of sand. At the same time the friction of all these grains of dust revolving round one another, disengages in a continuous manner real torrents of electricity. Above the whirlwind large birds of prey wheel in circles, either because they wish to enjoy the atmospheric equilibrium re-established by the storm, or because various small animals which are their food are carried along in the tourbillon.*

In mountainous countries the whirlwinds can raise neither clouds of animalcula nor masses of dust, but they carry into space those heaps of snow so terrible for travellers; more still, they remove even the pebbles and fragments of schist, gneiss, and granite, making them whirl in circles which move rapidly with the conflicting aerial currents. The geologist, Theobald, has seen some of these whirlwinds of stones which were no less than from 15 to 20 yards wide; it is not impossible, therefore, that certain masses of slaty fragments, which resemble piles raised by the hand of man, have been heaped up by whirlwinds.†

The marine whirlwinds being phenomena of the same nature as the terrestrial whirlwinds, must likewise raise particles from the surface that they traverse. The foam of the waves is sucked up by the aerial eddy, and ascends with a whirling movement. Sometimes the water swells and rises in great bubbles in the vacuum formed in the midst of the whirlwind by the attraction of the air towards its circumference. In spite of popular accounts, it is very rarely that the water is carried up to the low clouds which brood over the sea so as to fall in a deluge at a great distance, but the showers

* De Khanikoff, *Voyage dans le Khorassan*; Baddeley, Alexander Buchan, *Meteorology*.
† Theobald, *Jahrbuch des Schweizer Alpen-Club*, p. 531.

of salt water which are discharged far inland during hurricanes prove that this phenomenon is not impossible, and that entire masses of fluid, not merely vapours or scattered drops, can be drawn up into the kind of chimney which the storm makes. It is said that ships threatened by a waterspout have succeeded in destroying with cannon this moving column of vapours and in re-establishing the equilibrium of the atmosphere; but when the waterspout is of considerable dimensions, the passage of a bullet through the whirling vapours can only have but very passing results. Besides, a waterspout is rarely an isolated phenomenon; almost always it is connected with a tempest that the vessel cannot escape. Generally, too, the influence of the aerial eddies makes itself felt to a great distance from their apparent limits. Thus the masts of a ship have been broken by the wind when on the deck no violent movement of the atmosphere was perceived, and the tempest still seemed to be distant.

Unfortunately it must be said that the whirlwinds are of all meteoric phenomena those which are least carefully studied. Nevertheless it is certain that a profound knowledge of the various phenomena which occur in the formation of these slight aerial eddies, would enable us better to understand the grander cyclones, the entire system of the winds, and perhaps even the movements of the heavenly bodies, and the rotation of nebulæ. In the same way as embryology has contributed more than any other study to the development of anthropological science, so it is by following from the origin of its movement the particle of air which whirls in space, that we shall be able to explain in a clearer and more precise manner the great facts relative to the circulation of the air or even to that of the celestial bodies. While the astronomer burns to comprehend some prodigious cycle of the stars, too vast for his eye or his intellect, perhaps there exists under his eyes a simple whirl of leaves or dust, which he disdains even to look at, containing in its spirals the solution of the grand problem.

END OF SECTION I.

www.ingramcontent.com/pod-product-compliance
Lightning Source LLC
Chambersburg PA
CBHW030321240426
43673CB00040B/1233